Managing Distribution Retention Time to Improve Water Quality—Phase I

The mission of the Awwa Research Foundation (AwwaRF) is to advance the science of water to improve the quality of life. Funded primarily through annual subscription payments from over 1,000 utilities, consulting firms, and manufacturers in North America and abroad, AwwaRF sponsors research on all aspects of drinking water, including supply and resources, treatment, monitoring and analysis, distribution, management, and health effects.

From its headquarters in Denver, Colorado, the AwwaRF staff directs and supports the efforts of over 800 volunteers, who are the heart of the research program. These volunteers, serving on various boards and committees, use their expertise to select and monitor research studies to benefit the entire drinking water community.

Research findings are disseminated through a number of technology transfer activities, including research reports, conferences, videotape summaries, and periodicals.

Managing Distribution Retention Time to Improve Water Quality—Phase I

Prepared by:
Malcolm Brandt, **Jonathan Clement**, and **James Powell**
Black & Veatch
69 London Road, Redhill, Surrey RH1 1LQ, UK
and
Rob Casey, **David Holt**, **Neil Harris**, and **Chi Tuan Ta**
Thames Water Utilities

Sponsored by:
Awwa Research Foundation
6666 West Quincy Avenue, Denver, CO 80235-3098

Published by:

Awwa Research Foundation
Advancing the Science of Water®

IWA Publishing

DISCLAIMER

This study was funded by the Awwa Research Foundation (AwwaRF). AwwaRF assumes no responsibility for the content of the research study reported in this publication or for the opinions or statements of fact expressed in the report. The mention of trade names for commercial products does not represent or imply the approval or endorsement of AwwaRF. This report is presented solely for informational purposes.

Published by IWA Publishing, Alliance House, 12 Caxton Street, London SW1H 0QS, UK

Telephone: +44 (0) 20 7654 5500; Fax: +44 (0) 20 7654 5555; Email: publications@iwap.co.uk;
Web: **www.iwapublishing.com**

AwwaRF report number 91006F. Originally published by AwwaRF for its subscribers in 2003.
IWA Publishing version published 2005.

ISBN 1 84339 901 6

CONTENTS

TABLES

FIGURES

FOREWORD

The Awwa Research Foundation is a nonprofit corporation that is dedicated to the implementation of a research effort to help utilities respond to regulatory requirements and traditional high-priority concerns of the industry. The research agenda is developed through a process of consultation with subscribers and drinking water working professionals. Under the umbrella of a Strategic Research Plan, the Research Advisory Council prioritizes the suggested projects based upon current and future needs, applicability, and past work; the recommendations are forwarded to the Board of Trustees for final selection. The foundation also sponsors research projects through the unsolicited proposal process; the Collaborative Research, Research Applications, and Tailored Collaboration programs; and various joint research efforts with organizations such as the U.S. Environmental Protection Agency, the U.S. Bureau of Reclamation, and the Association of California Water Agencies.

This publication is a result of one of these sponsored studies, and it is hoped that its findings will be applied in communities throughout the world. The following report serves not only as a means of communicating the results of the water industry's centralized research program, but also as a tool to enlist the further support of the nonmember utilities and individuals.

Projects are managed closely from their inception to the final report by the foundation's staff and large cadre of volunteers who willingly contribute their time and expertise. The foundation serves a planning and management function and awards contracts to other institutions such as water utilities, universities, and engineering firms. The funding for this research effort comes primarily from the Subscription Program, through which water utilities subscribe to the research program and make an annual payment proportionate to the volume of water they deliver and consultants and manufacturers subscribe based on their annual billings. The program offers a cost-effective and fair method for funding research in the public interest.

A broad spectrum of water supply issues is addressed by the foundation's research agenda: resources, treatment and operations, distribution and storage, water quality and analysis, toxicology, economics, and management. The ultimate purpose of the coordinated effort is to assist water suppliers to provide the highest possible quality of water economically and reliably. The true benefits are realized when the results are implemented at the utility level. The foundation's trustees are pleased to offer this publication as a contribution toward that end.

Walter J. Bishop
Chair, Board of Trustees
Awwa Research Foundation

James F. Manwaring, P.E.
Executive Director
Awwa Research Foundation

ACKNOWLEDGMENTS

The authors are grateful to AwwaRF for funding this work and to the participating utilities listed below for their funding in kind, time, and survey response. Particular thanks go to the Project Advisory Committee (PAC) members for their time in reviewing the report and to the other survey participants.

Project Advisory Committee Members

- Stephanie Morales – AwwaRF Project Manager
- Robert Clark
- Dammika Vitanage
- John Proctor

Also thanks to Frank Oesterholt for providing information on the Dutch perspective.

Participating Utilities

- Champlain Water District
- City of Philadelphia Water Department
- Denver Water
- East Bay Municipal Utility District
- Elizabethtown Water Company
- Greater Cincinnati Water Works
- Greenville Water Supply
- Louisville Water Company
- South Central Connecticut Regional Water Authority

Other Survey Participants

- Amsterdam Water Supply, Netherlands
- Brisbane Water, Australia
- Duinwaterbedryf Zuid, Netherlands
- Melbourne Water, Australia
- Power & Water Authority, Darwin, NT, Australia
- South Australia Water Corporation, Australia
- South East Water Limited, Australia
- Sydney Water, Australia
- Thames Water Utilities, UK
- United Water International, Australia
- Vitens Friesland, Netherlands
- Vitens Overijssel BV, Netherlands
- WLN (Waterbedryl Groningen & WMD), Netherlands
- Yorkshire Water, UK

EXECUTIVE SUMMARY

When water leaves a treatment works and travels through a distribution system, its quality, with respect to many chemical and biological parameters, will degrade. The quality of the delivered water will be largely influenced by:

- The quality of treated water supplied into the network
- The condition of distribution assets within the network
- The retention time within the network

The water industry has focused predominantly on the quality of treated water and the physical condition of distribution assets when improving the quality of water at the customer's tap. However the quality of the water delivered is also affected by the time the water is retained in the different elements of the distribution network. Retention time is controlled both by the physical characteristics of the system and the operational regime. Physical characteristics such as pipe roughness may change throughout the life of the asset or be modified by rehabilitation. Operational activities may be structured; for example, pump scheduling and planned maintenance, or uncontrolled as in the case of demand driven operational responses.

Changes to water quality result from the reactions in the bulk water with time and through the chemical and biological reactions with the distribution system materials with which the water is in contact.

The chemical and biological reactions in the bulk water are relatively well understood enabling development of models and software for predicting changes. The reactions of the water at the pipe-water interface are more complex and less well understood.

RESEARCH OBJECTIVES

The aim of the research is to demonstrate that water quality within distribution networks can be managed effectively by controlling retention time and to develop practical and pragmatic methodologies for doing so. The project is divided into two phases.

Phase I focused on determining whether retention time can be used to manage the water quality by addressing the following core questions:

- Is water quality within a distribution system affected by the retention time of the water?
- Can retention time within a distribution system be determined?
- Can retention time within a distribution system be altered?

The review has concentrated on scientific theory, technical opinion and the operational experiences of international utilities. This Phase I report has addressed the three core questions and evaluated the current status of knowledge on:

- Parameters influenced by retention time
- Analysis tools and methodologies for determining retention time
- The analysis of retention time related water quality issues
- Engineering and operational solutions for reducing retention time

The objective of the project in Phase II is to develop, test, and validate practical and pragmatic methodologies, protocols, and analysis tools for managing water quality by controlling retention time. The tools and methodologies to be developed during Phase II have been identified through the work undertaken during Phase I and particularly during a workshop, where deficiencies in current tools and methodologies were highlighted. The participating utilities will test and validate the developments through field studies and practical application. This validation process will be used to refine the decision trees, methodologies, protocols, and analysis tools. The work completed in Phase II will be presented in a guidance manual that will compliment this report.

CONCLUSIONS FROM PHASE I

Water Quality Parameters

There is considerable information on water quality and the microbiological and chemical parameter changes that take place in distribution systems. However the available information can be unclear, site specific, and provide different conclusions. Difficulties for the water industry arise in finding appropriate practical and pragmatic solutions to distribution quality issues, as many of the factors involved are closely interdependent and action to control one issue may result in a different problem. The water quality parameters influenced by retention time, potential causes, and how they react are discussed and reviewed together with how the parameters are regulated worldwide.

Table ES.1 summarizes the water quality problems that are affected by retention time. The most strongly related parameters are:

- *Disinfectant residual and nitrification:* Decay of disinfectant residuals and nitrification are strongly related to retention time. The sensitivity of this relationship is highly dependent on the material and condition of the distribution pipes.
- *Disinfection by-products (DBPs), taste and odor, and microbial water quality:* These can be influenced by the disinfectant residual and so can be indirectly controlled by retention time.
- *Discoloration:* One of the causes of discoloration is stagnation and depletion of oxygen in unlined cast iron mains. This is particularly the case in the USA where mains are sized to meet much higher fire flow requirements than is common elsewhere. In such circumstances, discoloration is promoted by high retention time.

Table ES.1
Water quality problems affected by retention time

Problem	Parameters / measures	Potential causes
Microbial water quality	• Total Coliforms • Fecal Coliforms • HPCs*	• Reduction in disinfectant residual due to long retention time • Poor quality source water treatment (high AOC†/BDOC‡) • Intrusion • Unlined cast-iron pipe
Loss of disinfectant residual	• Chloramines • Free chlorine • Total chlorine	• Long retention time • High levels of TOC§ in finished water • Excessive amounts of unlined cast-iron pipe • Sediments
Disinfection by-products	• THMs** • HAAs††	• Long retention time • Inadequate pH control • Poor quality source water treatment
Nitrification	• Nitrite • Ammonia	• Excessive retention time causing release of ammonia • Microbiological activity • Poor quality source water • High organic content in source water
Discolored water	• Iron • Manganese • Turbidity • Color • Consumer complaints	*Red Water:* • Unlined cast-iron • Long retention time in unlined cast-iron pipe • Water chemistry characteristics (e.g. low alkalinity) • Change of flow regime *Sediment:* • Can be related to slow velocities (retention time) that result in deposition of particulates. Accumulated deposits can be swept up when velocity increases
Corrosion of metals	• Lead • Copper • Iron	• Inadequate chemical treatment (e.g., pH, carbonate, and orthophosphate) • Long standing times in pipe • Long retention time may reduce pH which could increase corrosion
Taste and odor	• Consumer complaints	• Chlorine residual • Phenolic compounds • Microbiological activity • Leaching from materials

* HPC Heterotrophic plate count
† AOC Assimilable organic carbon
‡ BDOC Biodegradable organic carbon
§ TOC Total organic carbon
** THM Trihalomethane
†† HAA Haloacetic acid

Analysis Tools

A considerable amount of research has been carried out, much of it under laboratory or pilot scale conditions, which has increased our understanding of the relationship between retention time and water quality changes. This has allowed the development of two approaches for predicting water quality changes, one based on calculating retention time and the second based on empirical knowledge and water quality data. These two approaches have led to the development of a number of tools for the analysis of water quality in distribution networks. Table ES.2 details the main tools available. Problems can be analyzed using simple manual calculations through to computer models and optimization techniques. Provided there are sufficient resources and computing facilities available, complex reactions and the interaction between different parameters can be modeled. However, any model requires levels of data and system information commensurate with the degree of accuracy of the required output. This generally leads to modeling assumptions which simplify the mechanisms and thereby maintain a balance between the availability of data and required accuracy of the predictive results. Furthermore, while some models have given good results, they are often still too complex to be applied as pragmatic operational tools and may require data which is not readily available to operators.

Control Methods

Many utilities have employed retention time management techniques without specifically classifying them as such. Table ES.3 summarizes the different methods by which retention time can be adjusted within distribution. Much can be done to reduce retention time by relatively simple and inexpensive techniques such as flushing at dead ends or reducing the operational top water level of reservoirs. Other techniques will involve more significant capital works such as downsizing of mains. Where it is appropriate, boundary valve changes to increase velocities or shorten flow paths, generally represent the least cost and the most effective solution. Flushing, while being a short-term measure, does not represent a long-term solution. While it is desirable to reduce retention time for water quality reasons, there may be other operational management policies such as leakage reduction which result in increased retention time. In addition, requirements for strategic storage for fire fighting and other emergencies conflict with high turnover and low retention times.

Feasibility Assessment Conclusions

The feasibility of various analysis tools and control measures was assessed by circulating a questionnaire to companies who were known to be interested in using this approach to manage water quality. Twenty-three utilities from USA, Australia, Denmark, UK, and the Netherlands returned completed questionnaires. The questionnaire identified the retention time related water quality parameters that were of greatest concern to the water industry, the tools and operational controls currently used to address those problems, and the feasibility of the analysis tools and control measures identified.

Table ES.2
Summary of retention time analysis tools

Method	Details
Manual calculations	Simple systems can be analyzed using manual and spreadsheet calculations. Pipe flow equations combined with storage or turnover assessments can be used to derive retention time estimates for snapshot conditions and simple system extended periods. Age is then compared with water quality records and customer complaints to identify parts of systems prone or liable to quality failure (chemical and biological).
Rules of thumb and experience	Generic rules are used together with hydraulic assessment and an understanding of system operations (e.g. for a specific Cl_2 dose, expect stagnant water complaints where age is over 100 hours). Operators' practical and specific experience can be used to identify parts of their system prone to water quality problems (e.g. under certain demand or supply conditions, problems may occur at known locations).
Tracer studies	By injecting a chemical into a network and then monitoring for it at a downstream point, the retention time can be directly measured. This technique is relatively straightforward, especially when the trace chemical is already present in the system and can be easily adjusted e.g. fluoride or chlorine.
Review of water quality data	Review of water quality data can indicate areas with water quality problems. These may then correlate with areas of high water age
Geographical information systems (GIS)	GIS can be used to select those properties which are at the extremities of the network or near dead ends and so vulnerable to a high water age. It is also possible to incorporate more sophisticated techniques by assigning characteristic water quality properties (including retention time) to each element in the GIS. Queries can then be set up which follow rules to estimate the water quality within the network. GIS can also be used to present database information spatially to look for patterns and causes of problems.
Hydraulic network model	Computer models of the hydraulic behavior of distribution networks such as EPANET, H2Onet, Stoner, WaterCad, Aquis, and Infoworks. This software can be used to predict the water age and highlight any areas in which it would exceed a given value.
Water quality network model	Many of the above network models can also be used to model water quality for parameters whose concentration is a direct function of water age. The most commonly modeled parameter is chlorine. However, this does require a reliable estimate of reaction coefficients that may not be readily available.
Water quality statistical model	Some utilities have developed models of water quality, which are purely derived from statistical analysis of the vast quantities of water quality data that are collected for regulatory purposes
Physical model	An expensive tool, not normally used for system operational management. However it may be useful for validating new design (e.g. the location of baffles to prevent dead zones in reservoirs).
Computational fluid dynamics (CFD)	CFD is increasingly being used to analyze the behavior of water within tanks and reservoirs. It can highlight dead zones in which there is limited turnover of water. Methods for rectifying problems such as installing baffles can also be modeled. CFD models have also been used to simulate flow in pipes including at dead ends under flushing conditions.
Reservoir model	Assessments of turnover including strategic storage, diurnal, and seasonal usage give indications of retention time and hence the preponderance of water quality problems. However three-dimensional modeling (CFD or physical models) may still be needed to understand internal circulation, dead zones, etc.
Customer perception surveys	Carefully designed and executed surveys can provide useful information and insight into particular problems. However, they need to be used in conjunction with other tools to provide information on retention time issues. They have been widely used in the UK in conjunction with structured sampling to identify and justify network improvement schemes.
Neural network models	Neural networks are potentially of benefit where it is reasonable to assume that there is a relationship between parameters, but where the mechanism by which they are related is not understood. They are effectively used to derive empirical relationships and could be used in a similar method to statistical water quality models (see above). To date they have not been widely used.
Optimization software	Optimization tools are widely available within spreadsheets (e.g. the 'Solver' add-in within Microsoft Excel) and can be used to optimize many problems that can be defined in terms of mathematical functions. Optimization tools are also increasingly being incorporated into other analysis software such as network models.

Table ES.3
Methods for controlling retention time

Method	Details
Altering the valving of networks	By changing valve arrangements and hydraulic boundaries, travel times can be reduced and water rerouted to increase velocities in low flowing pipes to maximize velocities
Installing time varying valves	Time and other control valves can be used to control flows
Manual flushing	Periodic flushing to remove sediment and reduce water age in dead ends and low velocity sections of pipe
Automated flushing	Automatic programmed flushing to remove sediment and reduce water age in dead ends and low velocity sections of pipe
Abandoning mains	Remove surplus capacity from system
Downsizing mains	Reduce system capacity to increase velocities and reduce retention times
Increasing turnover of storage tanks	Eliminate or reduce strategic storage and manage diurnal storage to minimize retention time. This is likely to be linked to resource capacity and operational optimization of both resources and pumping facilities
Adjusting pump schedules	Optimize pumping regime to match supply and demand and to minimize energy requirements and retention times (linked to optimizing system storage)
Taking storage out of service or removing tank cells	Remove surplus capacity from system
Reducing the TWL* of storage facilities	Eliminate or reduce strategic storage and manage diurnal storage to minimize retention time. The TWL may be varied with seasons.
Altering storage configuration to avoid dead zones	Replace common inlet outlet with separate pipes or install baffles to eliminate dead water zones thereby reducing retention time

* TWL = Top Water Level

The water quality parameters of greatest concern to the responding utilities were:

- Discoloration
- Disinfectant residual (often directly related with DBPs and nitrification)
- Microbial
- Taste and odor

The utilities recognized the importance and value of a wide variety of retention time tools for resolving water quality problems (Table ES.4). In all cases, greater than 50 % of the tools identified were considered viable for determining retention time and assisting with water quality issues. However, of the 14 tools listed only six were in common use by over 50% of the utilities. The tools that found most frequent application were:

- Review of water quality data
- Hydraulic modeling
- Rules of thumb
- Simple calculations

Table ES.4
Utility survey – usage and viability of retention time analysis tools

Tool	In Use	Considered Viable
Review water quality data	100%	100%
Hydraulic network model	95%	100%
Rules of thumb	90%	90%
Manual calculations	90%	80%
Customer perception surveys	65%	90%
Tracer studies	55%	70%
GIS	45%	80%
Reservoir model	45%	65%
Water quality network models	35%	85%
Physical model	35%	50%
CFD models	30%	60%
Water quality statistical model	15%	50%
Optimization software	5%	65%
Neural network models	5%	50%

It is of interest that the more complex and expensive options involving specialist software such as water quality network models, GIS, and CFD are considered viable but have not achieved widespread use. Water quality network models are used by just greater than a third of those who responded to the questionnaire.

Where retention time had been changed to control water quality many utilities had found this to be a successful approach. The most common methods for controlling retention time were manual flushing of pipe networks or methods for improving turnover in reservoirs. Mains flushing has been employed to manage all nine water quality parameters.

The Workshop

A Workshop took place over two days in November 2002 attended by around 20 delegates. The aims of the workshop were to:

- Disseminate the research findings
- Present case studies illustrating the use of retention time tools
- Discuss requirements of Phase II research

Case studies illustrating different tools for determining retention time and demonstrating benefits were presented and discussed at the Workshop. The case studies ranged from simple approaches for calculating retention time to those requiring detailed data analysis and the use of complex software. The case studies demonstrate that many utilities have utilized retention time tools to assist in planning changes to the distribution network and resolve difficult problems. They use the tools to identify where in the network problems occur and to solve a particular problem and they benefit from the experience gained to provide improved operational practice. Utilities that have used retention time tools for predictive purposes are generally seeking the optimum solution to a current problem, i.e. the tools are being used to assist with planning changes. In most case studies, the utilities have demonstrated benefits and this has encouraged them to use the same approach to solve similar problems and assist with deriving optimal solutions.

A number of important views were established from the Workshop:

1. There was unanimous agreement that water quality can be improved by managing retention time in distribution networks
2. The water quality problems for which retention time management might show most benefit were considered to be:
    - Chlorination and DBP formation
    - Chloramination and nitrification
    - Discoloration
3. Detailed field measurements should be undertaken before and after interventions in order to validate the effectiveness of improved retention time on water quality
4. Complex investigations involving hydraulic and water quality modeling have their role but generally after simple approaches have been exhausted
5. The guidance manual should be appropriate to both small and large utilities. It should take account of the different level of resources available to different utilities.
6. Deductive reasoning and common sense can solve many problems. It was therefore recommended that a question approach (list) to identify problems and solutions and for small or resource stretched operators be developed.
7. There is a need for simplified tools (rules of thumb) to manage retention time for operators to follow. This will encourage involvement in retention time issues and give ownership of the approach directly to the people operating the network.

The utility survey, case studies, and workshop discussions reinforce the conclusion that retention time management is potentially capable of controlling a number of water quality parameters. However there is a need to develop more effective guidance and compare the effectiveness of different techniques. Many tools are available for managing or improving water quality, ranging from simple spreadsheet calculations to complex and expensive software models. The latter are employed mainly by the large utilities for problem solving (reactive) rather than proactively during day to day operation of the networks. While the use of retention time tools is increasing in some utilities, there are many utilities in which they are not commonly used.

GUIDANCE MANUAL (PHASE II)

There is a clear need for comprehensive guidance that demonstrates a systematic and practical application of retention time analysis tools and solution options. This will be prepared during Phase II of this project and presented as a guidance manual. The guidance manual will provide a logical approach to analyze and solve water quality problems that are related to retention time. The manual will promote the use of deductive reasoning and common sense to solve a range of problems. It will be directly applicable to both small system operators and the large utilities.

An outline generic flow path has been developed that is relevant to many different water quality problems. The guidance manual will be based on a development of this flow path. It is not intended to produce prescriptive guidance, since each water quality problem is unique and there will be different resources available to each utility. However, the manual will be developed to incorporate further guidance and relevant information. Much of the knowledge contained in this report can be linked to the flow chart through a logical sequence of guidance notes and sequential instructions. To facilitate the linking of information, the guidance manual may be compiled as an electronic 'web style' document to facilitate the linking and layering of information.

CHAPTER 1
INTRODUCTION AND BACKGROUND

WHY IS RETENTION TIME IMPORTANT?

When water leaves a treatment works and travels through a distribution system, its quality, with respect to many chemical and biological parameters, will degrade. The quality of the delivered water will be largely influenced by:

- The quality of treated water supplied into the network
- The condition of distribution assets within the network
- Retention time within the network

To date utilities have focused predominantly on the quality of treated water and the physical condition of distribution assets when improving the quality of water at the customer's tap. However, the quality of the water delivered is also affected by the time the water is retained in the different elements of the distribution network. Retention time is controlled both by the physical characteristics of the system and its mode of operation. Physical characteristics such as pipe roughness may change over the life of the asset or be modified by rehabilitation. Operational activities may be structured; for example, pump scheduling and planned maintenance, or uncontrolled as in the case of demand driven operational responses.

Changes to water quality result from the reactions in the bulk water with time and through the chemical and biological reactions with the pipe wall and other surfaces in contact with the water. The mechanisms by which water quality degrades are very complex and will be affected by:

- The full chemical and biological make-up of the water, not just the concentrations of the individual parameters of concern
- The mixing of water from different sources
- Hydraulic characteristics
- Dispersion characteristics
- Temperature
- Time
- The material and condition of any surface in contact with the water, most notably the internal pipe wall surface area
- Physical and operational characteristics which result in contamination

The chemical and biological reactions in the bulk water are relatively well understood enabling development of software and models for predicting changes. The reactions of the water at the pipe-water interface are more complex and less well understood.

There is considerable information on water quality and the microbiological and chemical parameter changes that take place in distribution systems. However, the available information can be unclear, site specific, and conclusions may differ. Difficulties for utilities arise in finding appropriate practical and pragmatic solutions to distribution quality issues, as many of the factors involved are closely interdependent and an action to control one problem may result in a different problem.

Models have been produced to analyze many water quality parameters within distribution. These inevitably make assumptions to simplify the mechanisms. While some of these models have given good results, they are often too complex to be applied as pragmatic operational tools and may require data which is not readily available to operators.

Many utilities have implemented some forms of retention time management techniques without specifically classifying them as such. Some techniques can involve significant capital works, however much can be done to reduce retention time by relatively simple and inexpensive physical and operational changes such as flushing at dead ends or reducing the operational top water level of storage tanks, seasonally or permanently.

While it is desirable to reduce retention time for water quality reasons, there may be other operational management policies which can result in increased retention time:

- Leakage reduction
- Zoning for metering purposes
- Maximizing hydraulic capacity
- Provision of fire flows and strategic storage

Therefore a pragmatic approach will be required to balance conflicting objectives for the management of a system.

WHAT CONTROLS RETENTION TIME?

Fundamentally, retention time is determined by the flow rate and the internal volume of the distribution network. High flow, small diameter pipes, low storage, and short distances of supply will result in low retention times. Whereas low flows, large diameter pipes, high storage, and long distances of supply will result in high retention times. The above are determined by the design and operation of the network.

Network Design

Although design philosophies and criteria differ between utilities and countries, in general distribution systems are designed to supply the peak diurnal and seasonal demand for water from consumers while providing sufficient additional capacity for fire fighting purposes (fire flows). In practice, the sizing of the smaller pipes tends to be determined by minimum fire flow and pressure requirements. Some utilities also adopt standard minimum size pipes and pipe materials for use within the distribution network.

In Europe and many other international systems, per capita demand is generally significantly less than in the US as is the provision of flow and pressure for fire fighting (Twort, Ratnayaka, and Brandt 2000). These differences result in the minimum pipe sizes in the US commonly being 150 mm (6 in.) or 200 mm (8 in.) whereas in Europe fire hydrants are installed on 75 mm to 150 mm (3 in. to 6 in.) diameter mains.

Through a recent revision to the Dutch design criteria, in the Netherlands the minimum pipe size in new networks has been reduced in order to achieve a daily peak velocity of at least 0.4 m/s (1.31 ft/s) (van den Boomen and Vreeburg 1999). This velocity is considered sufficient to prevent sediment accumulating and produce a ‘self-cleansing’ network. The design approach allows for reduced fire flows to be agreed, taking into account the improved performance of modern fire fighting equipment. The approach results in new networks comprising of 40 mm

(1.5 in.) and 63 mm (2.5 in.) diameter pipes feeding domestic customers in local networks and dead ends where before the minimum size pipe may have been 100 or 150 mm (4 or 6 in.) diameter to cater for fire flows.

Retention time is relevant to the design of all pipelines. The majority of pipes designed for fire flows, when operating under normal demand conditions, will exhibit low velocities even under diurnal or seasonal peak demand conditions and hence the water will have a long retention time within the system. Conversely, systems designed to achieve self-cleansing velocities will also achieve a lower retention time of the water within the system.

Strategic storage within a system will have a direct impact on the age of water entering the downstream network. The amount of storage provided is generally related to the degree of risk and estimated water requirements to cover the duration of the emergency, such as fire, or failure, such as mains breaks. Long retention time within storage combined with poor circulation within the tank can increase the risk of excessively old water being introduced into a network.

Network Operation

The overall flow through a network is determined by the various 'demands' in the network. These include:

- Domestic and non-domestic consumers
- Non-revenue water usage
- Leakage
- Water abstracted for fire fighting

The flow in individual pipes and the distance of travel is controlled by the configuration of valves, the routing of flows, and the operational regime for managing sources, storage and pumping facilities.

The practice of subdividing networks into hydraulically discrete areas for leakage management, District Meter Areas (DMAs) risks causing high retention times in the vicinity of the boundary valves to the DMA. Designs in the UK of DMAs often now include an analysis of the age of water to assess the consequences of the additional closed boundaries and dead ends which are created. The initial design can then be modified to reduce excessive age and reduce the risk of old or stagnant water being supplied to customers.

DEFINING RETENTION TIME AND AGE OF WATER

In literature to date, the terms "retention time" and "age of water" have been used loosely and sometimes interchanged with each other when discussing distribution system water quality. However, it is important to differentiate between the two terms. Whereas age is a simple measure of time, retention time should always relate to a specific asset through which the water flows. The majority of systems have evolved over time, often having been extended extensively, reconfigured and renovated. Hence, water of the same age at a point of consumption can have been conveyed through pipes and facilities of different condition and characteristics in which case they are liable to have different water quality. For example, the water quality after twelve hours retention in an old unlined cast iron pipe is likely to be considerably worse than the water quality after twelve hours in a new clean storage tank. For this reason, retention time should always be considered in conjunction with the properties of the asset through which the water flows.

The following terminology and definitions have been used throughout this report:

Source: The source of a supply for a discrete hydraulic system is the last treatment process prior to the water entering supply. Generally the source will be the final storage at a treatment works, but it will include groundwater sources injected directly into supply and storage within the system where the water receives additional treatment prior to entering the zone downstream.

Age of water: Age of water at any point in the system is defined as the time a slug of water takes to travel from the source to a specific point of use. The age of water is related to the geometric characteristics of the pipes and network facilities and the velocities in each network component. In any network there will be multiple points where the age of water is the same. Over a period of time the age of water at a specific point will vary, both diurnally and seasonally, depending on the system demand for water. Age of water can be a measure of the overall "shelf life" of the water. Where age of water is calculated it can be used to assess where potential water quality problems are likely to manifest themselves based on guidance rules for the change in quality with time of a chemical or microbial parameter.

Blending: Blending can be due either to multiple sources feeding a discrete hydraulic system or to the multiple flow paths in a reticulated network. For both conditions, the age of water at any blend point is normally calculated as the flow weighted age of the different streams mixing at that point in the system. This definition should be treated with caution because for some water quality parameters it may be important to consider the age with respect to each source or the maximum age of the streams being blended.

Retention time: A limitation of using water age as an indicator of water quality is that the reaction rates are often controlled by contact with the internal surface of the distribution network. The geometry, material, and condition of those surfaces will vary throughout the network and so water with a given age in one part of a network may have significantly different quality to water with the same age in another part of the network. For example, the quality of water is likely to be significantly better where it has only passed thorough new medium density polyethylene (MDPE) pipes rather than where it has passed through a system which contains sections of old unlined cast-iron pipes. Retention time is defined in this report as the time it takes for a single slug of water to pass through an element of the network or group of network elements with similar physical characteristics. Retention time can be assessed for a single element, for example a storage tank, or for a combination of elements with the same or similar characteristics, for example unlined cast-iron pipe laid in a certain year or of a similar condition. Retention time is a more effective indicator of water quality than water age because it allows the time to be related to the condition of system components. Under specific analysis conditions, particularly where there is a shortage of detailed physical and condition data, retention time may be calculated for a network as a whole in which case age of water and retention time at any point within the system will be the same.

Velocity: One would normally expect the quality to deteriorate as retention time increases in an element, and hence low velocities would be detrimental to water quality. However, high velocities and changing velocities can also influence water quality. Higher velocities will encourage the erosion and entrainment of pipe wall deposits. The resultant higher turbulence (higher Reynolds number) may influence pipe surface processes and the rate of chemical mass transfer between the bulk flow and the pipe wall. Variations in velocity are of particular importance to discoloration. A regular high velocity may produce a 'self-cleansing' effect and purge the system of loose deposits before they accumulate in significant quantities. However, if higher flows only occur infrequently, then sediment is liable to accumulate and any subsequent high flow is liable to disturb those deposits resulting in discoloration.

AIMS AND OBJECTIVES

The aim of the project is to demonstrate whether water quality within distribution networks can be effectively managed by controlling retention time and to develop methodologies for doing so. The project is divided into two phases; the first phase focuses on the theory and the second phase focuses on its application.

Phase I

The objective of Phase I is to determine whether retention time can be used to manage the quality of water within the distribution system. This report presents the conclusions from this work which includes an international review of scientific theory, technical opinion, and operational experience. The work focuses on answering the following questions:

1. Is the water quality within a distribution system affected by the retention time of the water?
2. Can retention time within a distribution system be determined?
3. Can retention time be altered within a distribution system?

Phase II

During Phase II practical and pragmatic methodologies and protocols for managing water quality by controlling retention time will be developed and tested. The results of Phase II will be presented in a subsequent guidance manual.

CHAPTER 2
RETENTION TIME RELATED WATER QUALITY PROBLEMS

The rationale for this project is that the quality of water generally deteriorates with increased retention time. Table 2.1 lists the water quality problems which can be directly affected by retention time and summarizes the potential causes of the problems. These are discussed in this chapter in terms of how retention time affects water quality and how the parameters are regulated in different regions of the world.

MICROBIAL WATER QUALITY

Water entering distribution networks is not sterile but contains a natural mixture of bacteria which have little significance in terms of public health. Treatment of water ensures the absence of pathogenic microorganisms, primarily those that cause intestinal illness. Monitoring for the presence of specific pathogens in drinking water is impractical. Instead, evidence of fecal pollution from man or animals is determined by monitoring for the presence of indicator organisms. There are a number of different tests for microbial water quality:

- *Escherichia coli* is a natural inhabitant of the intestine of man and other warm-blooded animals or birds. If present in water it is likely to indicate fecal pollution.
- Coliform comprise many different species and are found widely in the environment. Their presence in water does not necessarily indicate fecal pollution. Research has shown that environmental coliforms can grow in distribution systems.
- Heterotrophic plate count (HPC) or standard plate count: This is a count of viable bacteria on selected agar media. It provides useful information on changes in the quality of water.

There have been many studies that have determined the effect of different parameters on microbial water quality within distribution systems, including:

- Measures of organic content primarily total organic carbon (TOC), biodegradable organic carbon (BDOC) or assimilable organic carbon (AOC)
- Temperature
- Retention time
- Dissolved oxygen
- Chlorine residual
- pH

The results of these studies demonstrate that many interrelated factors play an important role in controlling microbial water quality. The determining factor for deterioration of water quality with increasing retention time is whether the conditions within the network are conducive to the growth of the organisms of concern. High temperature, low disinfectant residual, high organic content, and a greater surface area to which microbes can adhere will promote microbial growth. A Dutch study simulating stagnation in customer service pipes invariably found exceptionally high plate counts (1600 cfu/mL) after two weeks stagnation (Spiering 2002).

Table 2.1
Water quality problems affected by retention time

Problem	Parameters / measures	Potential causes
Microbial water quality	• Total Coliforms • Fecal Coliforms • HPCs*	• Reduction in disinfectant residual due to long retention time • Poor quality source water treatment (high AOC†/BDOC‡) • Intrusion • Unlined cast-iron pipe
Loss of disinfectant residual	• Chloramines • Free chlorine • Total chlorine	• Long retention time • High levels of TOC§ in finished water • Excessive amounts of unlined cast-iron pipe • Sediments
Disinfection by-products	• THMs** • HAAs††	• Long retention time • Inadequate pH control • Poor quality source water treatment
Nitrification	• Nitrite • Ammonia	• Excessive retention time causing release of ammonia • Microbiological activity • Poor quality source water • High organic content in source water
Discolored water	• Iron • Manganese • Turbidity • Color • Consumer complaints	*Red Water:* • Unlined cast-iron • Long retention time in unlined cast-iron pipe • Water chemistry characteristics (e.g. low alkalinity) • Change of flow regime *Sediment:* • Can be related to slow velocities (retention time) that result in deposition of particulates. Accumulated deposits can be swept up when velocity increases
Corrosion of metals	• Lead • Copper • Iron	• Inadequate chemical treatment (e.g., pH, carbonate, and orthophosphate) • Long standing times in pipe • Long retention time may reduce pH which could increase corrosion
Taste and odor	• Consumer complaints	• Chlorine residual • Phenolic compounds • Microbiological activity • Leaching from materials

* HPC Heterotrophic plate count
† AOC Assimilable organic carbon
‡ BDOC Biodegradable organic carbon
§ TOC Total organic carbon
** THM Trihalomethane
†† HAA Haloacetic acid

Many studies have reported a correlation between coliform and heterotrophic bacterial counts and lack of disinfectant residual. Since retention time has a direct control on disinfectant residual, it is often assumed to indirectly control microbial activity. However, this is only valid where the disinfectant residual decays to a concentration which is favorable for microbiological growth. A number of models have been developed which assume microbial water quality to be a function of retention time due to this mechanism.

Microbial Water Quality Conclusions

Many factors affect the growth of microorganisms in distribution systems. If conditions are conducive to microbial growth then microbial counts will increase with retention time. In systems which maintain a residual disinfectant, an indirect but significant mechanism by which microbial water quality may deteriorate will be through the decay of that disinfectant with time.

LOSS OF DISINFECTANT RESIDUAL

In many countries, a disinfectant residual is maintained in treated water. The purpose of this is to protect against microbial contamination or regrowth within the network. By far the most common residual disinfectants are free chlorine or chloramine and it is well known that the residual concentration of both will decay with time. This is because chlorine and chloramine are strong oxidizing agents and react with many trace compounds within the bulk water or with material at the pipe wall. For a more extensive review of chlorine and chloramine decay the reader is referred to the AwwaRF reports *Characterization and Modeling of Chlorine Decay in Distribution Systems* (Vasconcelos et al. 1996) and *Chloramine Decomposition in Distribution Systems and Model Waters* (Valentine, Ozekin, and Vikesland 1998).

Free Chlorine

The decay of chlorine is frequently modeled as a first order reaction (see Chapter 3). However, there can be considerable variation in the rate of chlorine decay in different networks. First order decay coefficients have been widely reported mainly in the range of 0.01 to 10.0 hr^{-1} (0.24 to 240 d^{-1}) (Cardew et al. 2000, Powell et al. 1999, Vasconcelos et al. 1996) but occasionally fall outside this range. Figure 2.1 shows how this range of decay coefficients has a significant influence on the duration for which an acceptable residual remains in the water. This corresponds to variations in the time that chlorine decays to an unacceptably low level (Figure 2.1). Other models are discussed in the following chapter.

This variability is because the rate of decay is strongly affected by many different parameters which themselves show significant variation between distribution systems.

Temperature

The decay rate has been shown to increase significantly with temperature increasing two- to three-fold for every 50 °F (10 °C) rise in temperature. The Arrhenius equation has been shown by a number of researchers to give a good correlation between first order decay rates of bulk water with temperature.

Organic content

The decay rate has been shown to increase with organic content (TOC or UV absorbance). A linear relationship has been found by some researchers to give a reasonable correlation between first order decay rates of bulk water with temperature.

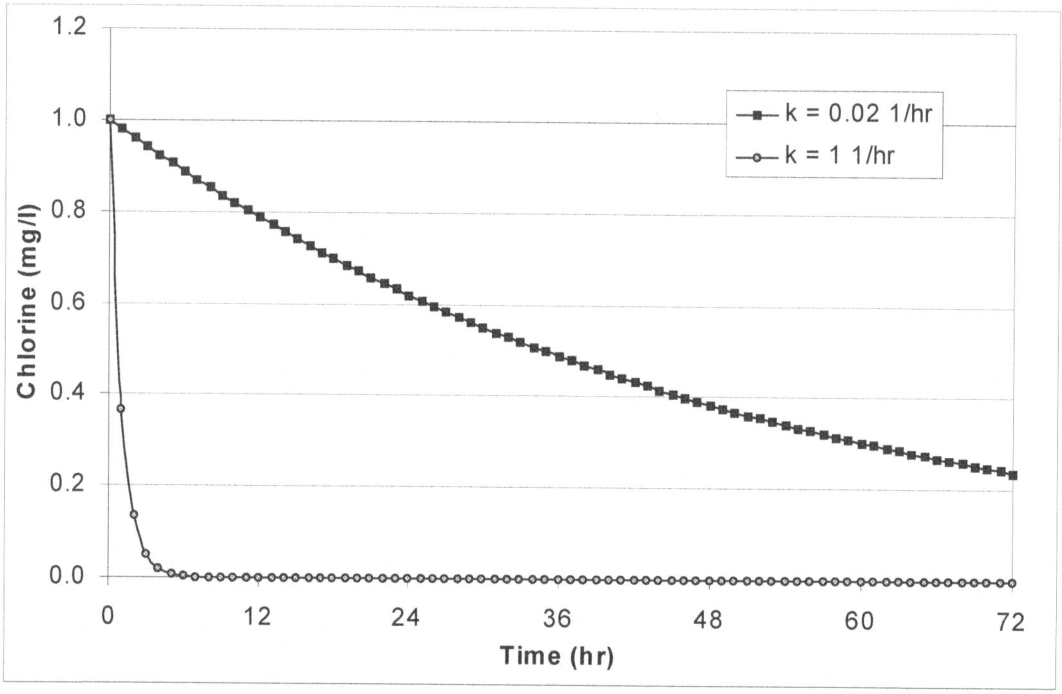

Figure 2.1 Illustration of first order decay with time

Re-chlorination

The decay rate invariably decreases when a sample is rechlorinated. This is probably due to the decay not being genuinely first order. The most readily oxidizable material will have reacted with the initial dose of chlorine and so upon re-chlorination, the concentration of readily oxidizable material will be less.

Pipe material

This has probably the biggest impact on the rate of chlorine decay. In order of reactivity (low to high):

- Polyethylene (negligible for new pipe)
- Polyvinyl chloride (PVC), (typical range 0.05 to 0.5 h^{-1} [1.2 to 12 d^{-1}])
- Epoxy lined pipe
- Cement or bitumen lined pipe

- Unlined iron pipes (typical range 1.0 to 10.0 hr^{-1} [24 to 240 d^{-1}])

Pipe diameter

It seems reasonable to assume that the rate of decay would be proportional to the ratio of internal surface to pipe volume i.e. inversely proportional to the pipe diameter. This is inherently assumed within the EPANET decay model.

Velocity

Some researchers (Woodward et al. 1996, Burgess et al. 1996, Powell 1998, Grayman et al. [N.d.]) have noted a tendency for increased rate of decay with increased velocity. The reason for this is not understood but it could be related to the mechanism by which chlorine transfers from the bulk flow to the pipe wall or due to the disturbance of sediment which subsequently reacts with the chlorine.

Chloramine

There has been less research into the decay of chloramine. Like free chlorine, it progressively decays with time, but the rate of decay is much lower. For this reason, chloramine may be used in preference to free chlorine in networks with a high chlorine demand, for example those containing a high proportion of unlined iron pipes. NanCIE (1998) showed that the decay due to reactions with pipe materials were lower for chloramine than for free chlorine by a factor of up to ten.

Some of the most extensive research into chloramine decay has been that undertaken by Valentine, Ozekin, and Vikesland (1998). They found that chloramine decay was primarily controlled by pH, initial chloramine concentration, inorganic carbon, and temperature.

Disinfectant Residual Conclusion

Irrespective of whether the disinfectant residual is in the form of free chlorine or monochloramine or which model is most appropriate, retention time is a fundamental factor which controls the amount of decay. The decay generally follows some form of curve with the rate of decay decreasing with time.

DISINFECTION BY-PRODUCT FORMATION

In distribution systems that use free chorine as a disinfectant, disinfection by-products (DBPs) can be a major concern (Kirmeyer et al. 2000). DBPs are formed when chlorine reacts with natural organic matter (NOM). The use of chloramine may also yield DBPs, but, with the exception of nitrites (see following section), are not discussed here due to insufficient published research. The DBPs of primary concern are trihalomethanes (THMs) and haloacetic acids (HAAs). These compounds are suspected carcinogens and as a result both Europe and the USA regulations impose tight limits on their concentration in drinking water. Other DBPs, most notably phenolic-based compounds, are of concern as they can cause taste and odor problems.

THM and HAA Formation

Early studies indicated that THM concentrations tended to increase in an exponential manner, asymptotic to a maximum formation potential (Clark et al. 1994). For some waters THM formation may be extremely rapid and therefore substantially complete within the water treatment process, but for other waters, THMs may continue to show significant increase over a period of days or even weeks (Vasconcelos et al. 1996, Joll et al. 2002). Reducing the time of contact between NOM and chlorine is often the key to reducing DBP formation. However, the rate of formation of DBPs depends on a number of factors (Singer 1999):

- The amount and nature of NOM: The chemical nature of NOM can vary significantly from water to water
- pH: The rate of THM formation generally increases with increasing pH whereas with HAA the rate increases with decreasing pH
- Temperature: As temperature increases the rate of formation increases
- The amount of disinfectant: Increasing the level of disinfectant will in general increase the rate of formation

To predict the level and rate of DBP formation a simulated distribution system (SDS) test can be conducted. This test is conducted by using finished water from a water treatment plant and then reproducing the chemical conditions in the distribution system. The tests are carried out over the time period in which water ages are expected to exist. Then by knowing the water age some prediction of the levels of DBPs can be made.

A limitation with the above test is that it does not include reactions of the pipe wall, which have been demonstrated to have a significant impact on THM formation. Valentine and Vikesland (2000) have shown that the interaction of the deposits from unlined cast-iron pipe can have a significant impact on formation of THMs. This interaction at the pipe-water interface may either increase or decrease the levels of THMs. HAAs have been shown to undergo biodegradation in the distribution system. These phenomena counter the traditional thought that the longest retention will yield the highest DBP levels.

Phenolic Based Compounds

Another important DBP issue is the formation of phenolic-based compounds that cause tastes and odors. A recent study (Khiari et al. 2001) has shown that homogenate anisoles that can be responsible for tastes and odors. The formation of these compounds is slow and so retention time will have a major impact on the formation of these compounds.

DBP Conclusion

The rate of DBP formation varies considerably depending on water characteristics and disinfectant concentration. In some circumstances, the formation reactions will be complete prior to the water entering the distribution network. In other circumstances, the formation may take a number of days, in which case their concentration will be directly related to retention time. However, interactions at the pipe wall are significant for some DBPs and this may counteract their formation.

NITRIFICATION

Chloramination is increasingly being used in preference to free chlorine as the residual disinfectant within distribution. This is particularly the case in the USA following the introduction of the DBP rule. Advantages of chloramination over free chlorine are that it decays much slower, is much less prone to THM formation, is better at controlling biofilm and higher doses can be applied without causing a chlorinous taste. The primary drawback of chloramination is nitrification.

Chloramines are formed by contacting chlorine and ammonia at a specific ratio. Over time chloramines breakdown and release free ammonia which is liable to be converted into nitrate by ammonia-oxidizing bacteria and subsequently into nitrite by nitrite-oxidizing bacteria. Problems associated with nitrification are (Kirmeyer et al. 1995):

- An increase in HPCs
- An increase in nitrite and nitrate concentrations (potentially harmful to infants)
- A decrease in pH potentially increasing corrosion
- A decrease in dissolved oxygen concentrations
- A reduction in the disinfectant residual

Temperature is a key factor in nitrification. As temperature rises, nitrifying bacteria become more active and convert ammonia to nitrite. In many systems nitrification is seen as a seasonal event, occurring mainly in the summer when temperatures are high.

Once nitrification starts it is very difficult to stop (Kirmeyer et al. 1993 and 1995). Monochloramine is relatively stable and may take many days to undergo significant decay. Therefore, areas with very long retention times are most prone to nitrification. Other researchers have shown a direct link with water age and nitrification. Methods for preventing nitrification involve primarily reducing the decay rate or reducing the total amount of decay. Such methods include:

- Increasing the pH: As pH increase over the range of 7 to 9 the chloramine decay rate decreases significantly.
- Removing unlined cast-iron pipe: Unlined cast-iron pipe causes rapid decomposition in chloramines.
- Removing NOM
- Reducing the water age or retention time
- Adjusting the chlorine to ammonia ratio

Nitrification Conclusions

Nitrification occurs in distribution networks by the conversion of free ammonia to nitrite by bacteria. Free ammonia can be present in the network either through excessive dosing or by the decay of chloramine. The latter is dependant on retention time, water quality characteristics and the condition of the distribution network.

DISCOLORED WATER

As water passes through some distribution systems it may accumulate particles that cause turbidity or color. This may be due to a number of causes (Marshall 2001):

1. Under conditions of low flow and low dissolved oxygen, unlined iron pipes may corrode and release soluble Fe^{II} into the water which may subsequently be oxidized to Fe^{III} precipitating as a red suspended floc. This mechanism was proposed by Kuch (1988) and has been the subject of much recent research in the USA (Sarin et al. 2001, Clement et al. 2002).
2. The disturbance of sediment or other loose material which has accumulated within the network. There are a number of different potential sources for this loose material which are discussed later. Discoloration caused in this manner has been the subject of much recent research in Europe (Smith et al. 1998 and 1999).
3. Oxidation of soluble Fe^{II} or Mn^{II} present in the source water into insoluble Fe^{III} (red precipitate) or Mn^{IV} (black precipitate). This may be a particular problem where there is blending of different source waters within a system as the sudden change in the chemical properties may trigger this oxidation (Elton, Brammer, and Tansley 1995). Since chlorine is an oxidant, it has also been suggested that the process could also be triggered by chlorination. This mechanism of discoloration is not directly affected by retention time and so is not considered further in this report.

The term red water is often used to describe iron based discoloration, whereas discoloration is a more general term. The subject of red water has been covered in depth by the AwwaRF report *Development of Red Water Control Strategies* (Clement et al. 2002) and is being further investigated in *Characterization of Processes Involved in the Generation of Colored Water in the Distribution System.* The former project concluded from survey data that in the USA, discolored water largely arises in unlined cast iron pipes. Clement et al. (2002) estimated that the total amount of unlined iron based piping materials within the United States was approximately 25 % and that in some older cities the amount of unlined cast-iron pipe may be as high as 75 %. Hence, there is a very high potential for elevated levels of iron in the water.

When water flows through unlined cast-iron discoloration can be caused due to the release of soluble iron from the pipe wall or due to the scouring of iron particles from the pipe wall (van den Hoven et al. 1994). Retention time plays a role in both mechanisms.

Red Water (Release of Soluble Iron)

Clement and his co-workers (2002) conducted experiments in the laboratory and with a series of 150 mm (6 in.) diameter unlined cast iron pipe loops. They found that the factors that have the strongest impact on red water are dissolved inorganic carbon (DIC), pH, dissolved oxygen, concentration of orthophosphate, and retention time.

DIC: The most important parameter controlling red water was shown to be DIC. As the DIC was increased from 2 to 15 mg C/L there was a dramatic decline in the amount of iron released. The effect of DIC on iron release was supported by field work with higher alkalinity (>60 mg $CaCO_3$/L) waters.

pH

Increasing pH from 7 to 9 decreased the iron release.

Dissolved oxygen

The authors established that as dissolved oxygen decreased, the amount of iron being released from exposed iron piping materials increased. Therefore, under slow flowing or quasi-stagnant conditions the amount of iron and red water could increase. The mechanism cited was that soluble reduced iron was able to move away from the pipe wall and into the bulk water stream. In the presence of dissolved oxygen the iron oxidizes to form a stable scale on the pipe wall.

Retention time

Reducing retention time by increasing water flow prevents the depletion of dissolved oxygen and maintains the pipe surface oxidized, thereby minimizing red water. In addition, the less time which a slug of water resides in contact with an exposed iron surface the less iron can be released into that water. This action is simple dissolution.

The above conclusions are based on work in the USA, but similar problems have been reported and investigated in Europe (Nielson 1994, Smith et al. 1998 and 1999). In the Netherlands, research into the cause of brown water in the city of Amsterdam concluded that the problem was partly caused by high retention time in small diameter unlined iron pipes. A linear relationship was observed between turbidity and retention time at several test sites.

Disturbance of Loose Material

In recent years, discoloration caused by the accumulation and then disturbance of sediment has been the focus of much interest in Europe. Accumulated material may be the result of:

- The deposition of sediment which has precipitated out within the network
- The deposition of sediment contained in the source water
- The corrosion or deterioration of the internal surface of pipes
- Biological growth on the internal surface of pipes
- Ingress of material during the construction or maintenance of the network

Material will accumulate under low flow conditions most likely when the retention time is at its greatest. The build up of sediment may be a slow gradual process. However, once sediment has accumulated there is a major risk that any abnormally high flow will rapidly entrain sediment and cause discoloration.

The threshold velocities at which sediment is deposited and entrained will vary with: the pipe diameter and roughness; the sediment size, shape and density; and any cohesive forces holding the material to the pipe wall. Sediment transport in marine and industrial environments has been widely researched and a number of validated models have been developed. However, the conditions within water distribution systems are significantly different; the sediment is much smaller, its concentration is much lower and various chemical and biological bonds may adhere it to the pipe wall. This means that the applicability of most existing sediment models to water distribution networks is highly questionable. Nevertheless, such models have been used as the basis for determining where deposits will occur, design of self-cleansing systems and calculating ideal flushing velocities to remove deposits. Ackers, Brandt and Powell (2001) selected models which they deemed to be most appropriate to the conditions within distribution systems and used them to calculate a set of guidance velocities for various pipe properties.

Discolored Water Conclusions

Changing the retention time in a network is often synonymous with changing the flow velocity and so can have a direct impact on the risk of discoloration. Conditions of low or zero flow (high retention time) favor the release of soluble iron from pipe walls. Low flows also promote the deposition and accumulation of sediment with the risk that any subsequent higher flows may disturb that sediment and cause discoloration.

CORROSION OF METALS

While the dissolution of lead and copper into water supplies is related to retention time, lead and copper pipes are predominantly found in domestic and industrial premises rather than the utilities distribution network. Hence control methods for lead and copper are focused at the treatment works rather than the retention time within consumers' premises over which utilities have little control. Lead and copper are therefore not considered further in this report.

TASTE AND ODOR

Taste and odor and discoloration are generally the two major causes of customer complaint. This is because they are the water quality measures which are most immediately perceptible to customers. In 2001 over 30 % of water quality complaints in the UK were associated with taste and odor (DWI 2002).

There are numerous different chemicals which can potentially cause taste or odor problems in drinking water and some may be due to a mixture of different chemicals. These chemicals may be derived from a number of sources including their formation in the distribution network where retention time is an important factor:

- Disinfectants
- DBPs
- Natural organic contaminants in the source water, in particular blue green algae or *actinomycetes*
- Pesticides in the raw water
- Microbial activity within distribution networks: Most potent taste & odor producing chemicals are microbial metabolites (Young et al. 1996).
- Reactions occurring within distribution
- Reactions occurring at domestic properties
- Blending of different water sources: Although not perceptible in either of the individual waters, when they are blended it is possible that a reaction will occur or a mixture of chemicals be formed which could cause a problem.

- Under stagnant conditions sulfate reducing bacteria can reduce sulfate to sulfide which may result in taste and odor problems

Chlorine is one of the most commonly cited causes of taste or odor. In 2001 over 50 % of all taste complaints in the UK were related to chlorine residual (DWI 2002). An early driver for much of the UK interest in chlorine network models was to investigate methods of configuring networks so as to minimize the chlorine dose while still retaining a measurable residual at all points within the network (Brammer and Tansley 1993).

Taste and odor problems are often difficult to resolve as their cause can be complicated by a number of factors:

- Chemicals may cause taste or odor at well below the prescribed concentration for that chemical or even below the detectable limit for that chemical
- The nature of a taste or odor may vary with concentration
- Some tastes and odors may be masked by the stronger taste of a disinfectant residual. The chemicals causing an earthy taste may actually be present in a source water, but only become apparent towards the limits of a network where the residual disinfectant has decayed.
- The perception limit for tastes will vary considerably between different people
- Chlorinous tastes may not be directly caused by the disinfectant residual, they could be caused by DBPs, such as chlorophenol
- Constant taste or odor is much less likely to cause a complaint than changes in taste or odor (MacRae, Keay and Farrimond 1993)
- Sensory response is generally logarithmic (MacRae, Keay and Farrimond 1993). Hence a considerable change is often required in the concentration of a taste-causing chemical to make any appreciable difference in the taste or odor.

Taste and Odor Conclusions

While there are many different causes of taste and odor, only some of them will be directly affected by retention time. Those which are partly affected by retention time include:

- Disinfectant residuals
- Tastes which are masked by a disinfectant residual and only become apparent once that disinfectant has decayed
- Those caused by microbial activity within the distribution network

While controlling retention time may be of some benefit in controlling the taste and odor of water, it will often be more effective to improve treatment processes to ensure efficient removal of the precursors. Furthermore, taste and odor is subjective and is often influenced by factors other than water quality and retention time, for example receipt of a bill from a utility.

INTERNATIONAL REGULATIONS

In many countries standards and regulation of drinking water parameters are based primarily on the World Health Organization (WHO) guidelines. WHO published their second edition of the Guidelines for drinking-water quality in three volumes between 1993 and 1997.

The volumes are Volume 1, Recommendations (WHO 1993), Volume 2, Health criteria and other supporting information (WHO 1996), and Volume 3, Surveillance and control of community supplies (WHO 1997). The primary aim of the Guidelines is the protection of public health. They are intended to be used as a basis for the development of national standards that, if properly implemented, will ensure the safety of drinking-water supplies through the elimination, or reduction, of constituents of water that are known to be hazardous to health. The guideline values recommended are not mandatory limits. In order to define mandatory limits, it is necessary to consider the guideline values in the context of local or national environmental, social, economic, and cultural conditions.

In the USA, the United States Environmental Protection Agency (USEPA) set limits for water quality parameters through promulgating regulations and these are then legislated for in the individual states. Each state can adopt its own prescribed value for each water quality parameter provided that it is not less stringent than the primary regulation. The European Union procedure is similar to the USA in that member states are required to transpose Drinking Water Directives issued by the Commission into national legislation. In Japan, the 'Waterworks Law' specifies limits for a range of water quality parameters related to public health and the acceptability of water to consumers. However, Japanese water utilities generally operate to more stringent guideline values. Other countries such as Australia and New Zealand have produced guidelines (Anon 1996), not mandatory standards, for drinking water quality parameters. These are mainly based on the WHO guidelines and any departures are detailed in the text of the individual country guidelines.

Since many countries have adopted the WHO guidance on water quality, goals relating to health based contaminants are therefore similar throughout the world. Differences are due to, adoption of either more stringent or more relaxed criteria through a realization that there are greater priorities in water supply at the local or national level. Therefore those parameters where retention time plays a role in water quality change are similar worldwide.

Table 2.2 and the following sections identify those water quality parameters that may change as the age of water increases.

Table 2.2
Summary of regulations for finished water quality

Parameter	Sample location	Regulatory limit	Reference	Comments
Disinfectant residual	Entry point to distribution system	0.2 mg/L on a continuous basis	SWTR*	Only applies to systems using surface water supplies
		None	UK std†	No UK limits on disinfectant residual. Residual must be sufficient to ensure appropriate bacteriological quality
	Customer' taps	0.1 mg/L free chlorine	JP std‡‡	Guideline maximum of 1.0 mg/L
Disinfectant residual or HPC bacteria count	Throughout distribution system	Detectable level of disinfectant residual or HPC bacteria count of 500 or less cfu/mL in 95 % of samples collected monthly for any two consecutive months	SWTR	Only applies to systems using surface water supplies. In the United States, *Legionella* is also regulated by a treatment technique
	Throughout distribution system	No abnormal change	UK std	UK storage facilities are required to be sampled weekly for HPC
		100 cfu/mL	JP std	
Nitrite	Entry point to distribution system	1.0 mg/L as N	NPDWR‡	Maximum contaminant level
	Entry point to distribution system	0.1 mg/L as N	EU std§	Maximum admissible concentration
	Customer' taps	0.5 mg/L as NO_2	UK std	Monitoring only required when chloramination is practiced
	Throughout distribution system	10 mg/L as N	JP std	Guideline value 0.05 mg/L as N
TTHMs	Throughout distribution system	100 µg/L, running annual average of quarterly samples	DBPR**	Systems serving > 10,000 people
	Customer' taps	100 µg/L	EU std	The specified compounds are : chloroform, bromoform, dibromochloromethane, bromodichloromethane
		100 µg/L	JP std	Lower values for specific THMs
Total coliform bacteria	Throughout distribution system	0 cfu in 95 % of samples	TCR††	Number of samples determined by population served
	Treatment works and customers' taps	0	UK std	
	Storage facility	0 cfu in 95 % of samples	UK std	UK storage facilities are required to sample weekly
		0	JP std	

Source: Adapted from Kirmeyer et al. 1999
* SWTR – Surface Water Treatment Rule
† UK std – United Kingdom standard regulations
‡ NPDWR – National primary drinking water regulations
§ EU std – European Union standard regulations
** DBPR – Disinfectant and Disinfectant By-Products Rule
†† TCR – Total Coliform Rule
‡‡ JP std – Japanese standard regulations. Note Japanese utilities will generally operate to much stricter guideline values.

Disinfectant Residuals

In the USA, the current Surface Water Treatment Rule (SWTR) has numerous requirements for systems treating surface water. Part of the rule applies to the disinfection of distributions systems. It requires that all systems treating surface water provide a continuous disinfection residual of at least 0.2 mg/L entering the distribution system. The system must demonstrate a measurable disinfectant residual or a heterotrophic plate count of less 500 cfu/mL in at least 95 % of the samples. The samples are collected at the same frequency as coliform samples.

By contrast European Union legislation allows member countries the freedom to choose whether or not they maintain a residual disinfectant in their networks. This is most pronounced in the Netherlands where many water companies do not maintain a residual.

Coliforms

The Total Coliform Rule (TCR) in the USA requires all public water systems to analyze distribution system water samples for total coliform bacteria. The number of samples required is based on population density or the number of service connections. For systems required to collect 40 or more samples a month coliforms must be absent in 95 % of the samples. For systems collecting fewer than 20 samples monthly then coliform bacteria must be absent in all but one sample. Any detection of coliform bacteria must be reported to the State. Samples that test positive for coliforms must be analyzed for fecal coliforms or *E. coli*.

The TCR may be revised with particular attention to retention time and determination of representative sampling locations. Many water systems select sampling sites that may have low water ages where flow is high and coliforms are likely to be absent. The TCR does not include specific requirements to collect and analyze water samples from storage facility and their outlets.

The recent revision of the European Directive (98/83/EC) on the quality of water intended for human consumption provided an option for coliforms to be classified as an indicator of the quality of water and not necessarily fecal contamination. This option was not adopted in all member countries including the UK. In the UK there are requirements to monitor for coliforms in treatment works, service storage facilities, and at customers taps. The annual number of samples for analysis is determined from the volume of water supplied. To meet compliance, no coliforms should be present in 100 mL samples of water leaving treatment works and 95 % of samples (100 mL) for coliforms from storage facilities are negative.

Disinfection By-Products

THMs and HAAs are regulated in the USA by the Stage 2 Disinfectant and Disinfectant By-Product Rule (DBPR). The MCLs are 80 µg/L for THMs and 60 µg/L for HAAs. Fixed sample sites are used and the 'Locational Running Annual Average' (LRAA) for each site must comply with the MCLs. An Initial Distribution System Evaluation (IDSE) is under consideration. If implemented, the IDSE will require utilities to select representative sample sites, unless the water utility can demonstrate from two years of data that distribution samples are less than 40 µg/l for THMs and 30 µg/l for HAAs. There are two options for conducting an IDSE:

1. Conduct a Standard Monitoring Program (SMP). This involves an increased level of THM and HAA monitoring for one year. The data is then reviewed to determine representative sites for compliance monitoring.
2. Alternatively the utility can conduct a System Specific Study (SSS). This allows a reduced level of sampling by utilizing specific knowledge of the system to substantiate the selection of representative sites. In particular the modeling of retention time is often a key element of an SSS.

The recent European Directive (98/83/EC) specifies a maximum concentration of 100 μg/L for THMs, as measured in samples taken at the customers' taps. The specified THM compounds are, chloroform, bromoform, dibromochloromethane and bromodichloromethane. No specific regulatory limits are given for HAAs. These regulations appear less stringent than those promulgated by the USEPA.

Nitrite

There are significant differences in nitrite standards between countries. In the USA the standard is set at a maximum concentration of 1.0 mg/L as N entering the distribution system. The new EU standard is 0.1 mg/L entering distribution and 0.5 mg/L as N at consumers' taps. This is considered less stringent than the standard set in the earlier EU Directive. In addition, for the water to comply with wholesomeness, it must also satisfy the formula:

$$\frac{[Nitrate]}{50} + \frac{[Nitrite]}{3} < 1 \tag{2.1}$$

where $[Nitrate]$ = nitrate (NO_3) concentration (mg/L)
$[Nitrite]$ = nitrite (NO_2) concentration (mg/L)

pH

pH requirements in many countries are set to limit corrosion of metals in distribution networks, particularly with respect to lead and copper. Maintaining a pH just above neutral (7.5 to 8.0) both prevents corrosion and allows some residual disinfectant activity from chlorine and chloramines. In the USA there is a minimum pH guideline of 7.0 whereas in Europe the pH standard is 6.5 to 10.0. In distribution networks with cement lined pipes the pH can rise above 10 through leaching from the lining if the treated water is not sufficiently buffered, such as low alkalinity waters.

CONCLUSIONS

Loss of disinfectant residual and nitrification of chloraminated waters is strongly controlled by retention time. A number of other parameters are affected by the disinfectant residual and so can be indirectly controlled by retention time. These include:

- *DBPs*: These are a by-product of the decaying disinfectant residual and so may be expected to increase with retention time. However, this is not always the case either due to subsequent biological and chemical reactions which can break down some

DBPs (e.g. HAAs) or because they can form very rapidly and so their formation may be substantially complete upon leaving the treatment works and show no significant further growth within distribution.

- *Taste and Odor*: High chlorine concentrations can mask other less agreeable tastes and odors in a water which are revealed once the chlorine has decayed. Consumers may also be sensitive to fluctuations in the concentration of the disinfectant residual.
- *Microbial*: Although there is inconclusive direct evidence of a correlation between retention time and microbial water quality, once a disinfectant residual decays, the water will be more vulnerable to microbial contamination or regrowth.

International differences in regulation and practice result in differences in the significance of these parameters and the degree to which they are affected by retention time. The higher chlorine concentrations common in the USA compared to Europe mean that some of the above indirect relationships may be more apparent in the USA than Europe.

Discoloration can also be affected by retention time, but again there is an international difference in perspective. In the USA, discoloration is most commonly seen as being due to stagnation and oxygen depletion in unlined cast iron mains. Whereas in Europe disturbance of sediment deposits when there is an increase in flows is also seen as a common cause. This difference is partly due to the higher fire flow requirements in the USA which typically result in significantly larger mains than would be used in Europe.

CHAPTER 3
INVESTIGATING RETENTION TIME RELATED WATER QUALITY PROBLEMS

This chapter identifies the various tools that are available for evaluating retention time and estimating its impact on water quality. The tools are based on a variety of different types of technique: field exercises, laboratory studies, general rules, manual calculations, computer modeling, physical modeling, data visualization, and data analysis. Some tools are already in wide use within the water industry whereas others are still at a development stage. The chapter is divided into four sections, concluding with a summary of the available tools.

Tools for estimating retention time: For most retention time management techniques, it is necessary to estimate the retention time for a given network configuration. For some purposes, it may be appropriate to predict theoretical retention times corresponding to various alternative network configurations.

Tools for assessing how water quality changes in relation to retention time: Estimating retention time for current and alternative network configurations will allow operators to see whether it is possible to reduce the retention time significantly. However, it does not tell us whether a given retention time is likely to result in acceptable or unacceptable water quality. Hence tools are also needed to assess what the maximum acceptable retention time would be for a particular system.

Tools for predicting water quality changes based on retention time: It may be appropriate to take the analysis a step further and predict the behavior of different water quality parameters. Most, but not all of these techniques involve predicting retention time as part of a more detailed modeling exercise.

Tools for predicting water quality changes that do not require calculating retention time: Some tools are based on an understanding of the factors that affect water quality. These do not require calculation of retention time but relate to associations of key parameters that affect water quality; for example temperature and AOC.

Chapter 6 discusses how to select appropriate tools for a given problem.

TOOLS FOR ESTIMATING RETENTION TIME

Estimating retention time does not necessarily require sophisticated hydraulic modeling techniques. In some circumstances they may be appropriate, but this will not always be the case.

Manual Calculations

Manual calculations offer a quick and easy method of estimating retention time and are commonly used to examine single trunk mains with simple flow patterns. The formulae for manual calculations are based on the following simple mass continuity equations for pipes:

$$t = \frac{L}{u} \tag{3.1}$$

or

$$t = \frac{V}{Q} \tag{3.2}$$

where

$$Q = uA \tag{3.3}$$

$$V = LA \tag{3.4}$$

$$A = \frac{\pi D^2}{4} \tag{3.5}$$

t = time
L = length
u = velocity
V = volume
Q = flow
A = cross-sectional area
D = internal diameter

Although simple and constrained to single pipes, manual calculations have a wide application for common network issues such as:

- Estimating the minimum flow to avoid excessive retention time in major trunk mains
- Estimating minimum pump running times on pumping mains to storage facilities

An advantage of manual calculations is that they can be used when a network model is not available. Also, they can be used widely by operatives and technicians who may not be able to operate a more complex model for estimating retention time.

The minimum retention time will vary considerably depending on the water chemistry and pipe materials. The estimate used in manual calculations is often based on experience and local knowledge of water quality and pipe conditions.

Spreadsheet Models

The simple calculations above can be taken a step further by using spreadsheet models which account for time varying diurnal flow patterns. The increased use of telemetry for remote monitoring of distribution networks has led to an increased availability of flow and tank level data which can be quickly copied into spreadsheets. The retention time or rate of turnover in mains can be estimated by calculating the volume of water entering the main for each time interval of flow data and comparing this with the internal volume of the main. In a similar manner historic records of tank level, input flow, and output flow can be used to provide an initial assessment of storage retention times and hence those tanks likely to have turnover issues.

This type of analysis is only suitable for storage facilities with short inlet and outlet mains. Once the storage in the inlet or outlet main becomes significant it needs to be modeled together with the tank for which a network model is often a better tool. Simple spreadsheet models normally assume perfect mixing within the tank which is often not the case. To analyze mixing within the reservoir, tools such as computational fluid dynamics (CFD) (discussed later) or compartment models should be used as described in the AwwaRF report *Water quality modeling of distribution system storage facilities* (Grayman et al. 2000). Nevertheless, for water systems where telemetry data is available, spreadsheet models of retention time are a useful means of quickly surveying key assets for potential quality issues.

Computer Distribution Network Models

Network models attempt to define the hydraulic behavior of distribution networks by mathematical equations to be solved by computers. If accurate, they will allow the user to investigate the behavior at any point in the network without having to go into the field and measure it directly. In addition, they allow the user to consider what would happen if different conditions were imposed on the network; for example, to investigate whether a network is likely to cope with the increased demand for water from a new housing estate. Network models used to predict hydraulic behavior are widely adopted within the American and European water industries. Over the last fifteen years there has been interest in the use of network models to model water quality parameters and in particular chlorine. These models are discussed in a subsequent section of this chapter. There are numerous software packages available for building network models. The software package EPANET, which was developed by the USEPA, is frequently used by researchers as a platform for further development. EPANET is in the public domain and can be downloaded for free from the internet (USEPA 2002).

Solution Techniques

Network models generally represent distribution networks as a set of nodes connected by links. Normally a link will be a stretch of pipes with similar properties (diameter and roughness). Depending on the software package components such as pumps, tanks, or valves may be either classified as nodes or links, but this makes little practical difference to the use of the model. The properties for every component in the model must be defined, so that the network behavior can be mathematically calculated. These parameters would include:

- Pipe length, diameter, and roughness
- Node elevation
- Demand (flow abstracted) at each node

The hydraulic analysis is invariably based on a technique developed in 1935 by Professor Hardy Cross (cited in McGhee 1991). The basis of the technique is:

1. The sum of flows into a node equals the sum of flows out of a node. This is an application of the principle of conservation of mass.
2. In any loop, the sum of all head losses is zero.

The flows within each loop are adjusted iteratively until this second condition is satisfied. There is some variation between models in the way this is done (Brendan and Kuczera 1997).

To model a system over a period of time the above procedure is used, but it is applied successively, typically at a time interval of one hour. This results in a 'sequential steady state' rather than a true dynamic model. These models ignore the continuous variations in pressure and flow which occur within each time step and ignore transient pressure waves (surge or water hammer) which are much more complex to model.

Network models invariably produce a large amount of output data. Most modern modeling packages incorporate a graphical representation of the network which displays output data and can be used to edit the properties of individual nodes or links (Deininger et al. 1992). Figure 3.1 shows the user interface for the EPANET modeling package. Model results can generally be viewed via annotated maps, tables, graphs, color coded plots or contour plots.

Age of Water (Water Quality) Models

Many hydraulic network modeling packages incorporate algorithms to calculate the age of water at any point in the network (Liou and Kroon 1987, Boulos et al. 1994, Boulos et al. 1995, Rossman et al. 1993, Reddy et al. 1995). Models that are used to predict water age are frequently referred to incorrectly as "water quality models". Nevertheless, calculating water age is a fundamental first step when using models to predict water quality.

Unlike flow and pressure, which are only affected by the current conditions imposed on the network, the age or water quality of any parcel of water will depend on the integrated effect of the flows experienced by that parcel during preceding time-steps since entering the network. Hence, water quality models have to trace individual parcels of water as they pass through the system. This requires the model to perform the analysis at a much greater temporal and spatial resolution than is necessary for hydraulic models, thereby greatly increasing the computation required. There are various methods by which age can be calculated. Work by Rossman and Boulos (1996) compared the commonly used methods and concluded that while some were slightly more efficient than others, their numerical accuracy was very similar.

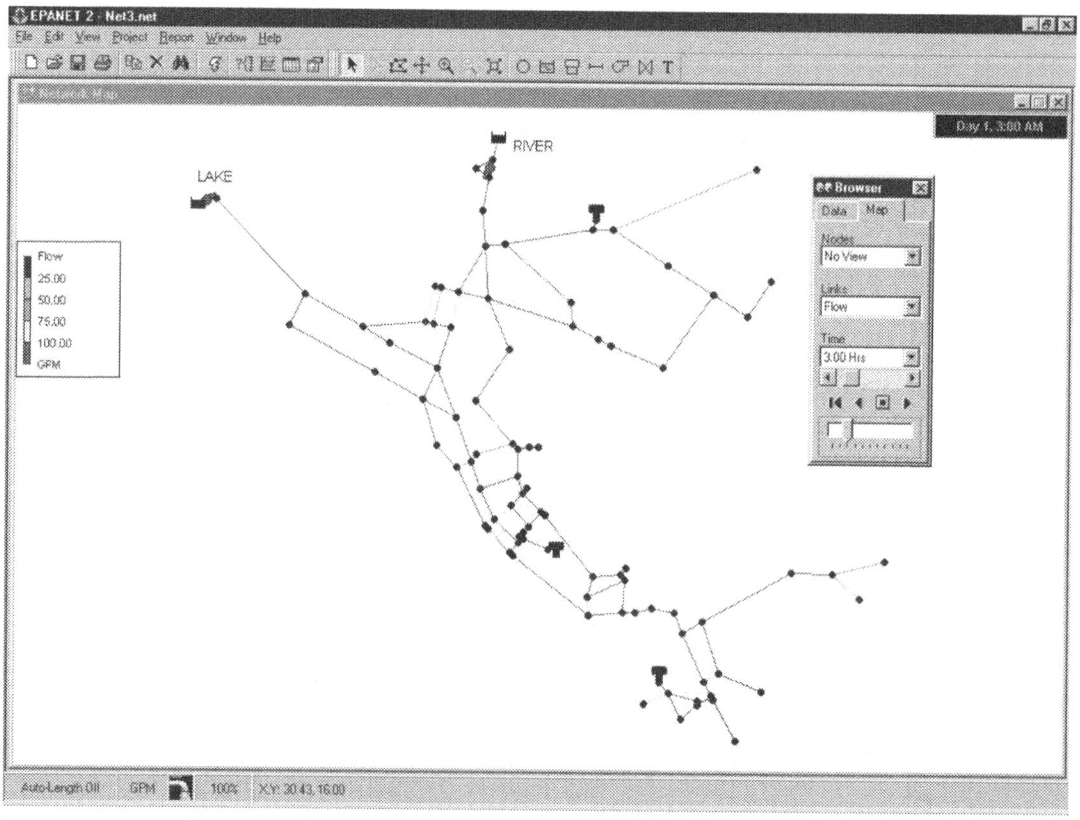

Figure 3.1 EPANET graphical display

Model Build

The model building process involves defining the value of the input variables for the model. The variables represent the properties of all the components within the network. If the properties are unknown, then assumptions need to be made for estimating their value in order for the software to calculate a solution. The accuracy of these estimates is the fundamental factor which determines the accuracy of the final model. Hence, the modeler needs to keep in mind the accuracy of the output required from the model and quality and availability of the data from which the model is being built. Successful construction of a validated model requires a sound knowledge of water distribution, a methodical approach to both model construction and the fieldwork activities, development of practical procedures to simplify future model updating and a systematic approach during the calibration process (Twort, Ratnayaka, and Brandt 2000).

The increasing use of geographic information systems (GIS), customer information systems, and associated system performance databases, provide opportunities for building and updating hydraulic models electronically thereby eliminating the costly manual model construction process which has inhibited construction programs in the past. Further developments with optimization methods should bring additional construction cost savings and analysis facilities.

This report does not give full guidance on the model build process. The aim in the following sections is to explain the benefits and limitations of using these models to analyze retention time related water quality problems. For more general guidance, the reader is referred to the books by Walski, Chase, and Savic (2001); Walski et al. (2003); Clark and Grayman (1998); and WRc (1989).

Model Calibration and Verification

At the end of the initial model build, the user should have a model which will solve numerically. However, the initial model is based on assumed values for the many parameters which are not directly measurable, such as pipe roughness or certain water quality coefficients. Until the model has been calibrated, there can be little confidence in its output. Calibration is achieved by comparing model predictions with field measurements and adjusting the value of model parameters to improve the fit between predicted and observed data. The parameters adjusted during the calibration should be those in which there is greatest uncertainty and so an understanding of potential errors is important (Engineering Computer Applications Committee 1999). The parameters most commonly adjusted during hydraulic calibration are the pipe roughness valve status (degree of throttle) and demand characteristics. Pipe diameter can also be adjusted, although this is less common, head-loss is very sensitive to diameter and small errors in actual diameter can lead to significant errors in head-loss predictions. It is fairly common to use nominal diameters for pipe size and these can be significantly different from actual diameters, especially if the pipe is lined or has corroded.

The calibration process effectively checks whether the model accurately predicts the observed head-loss between measurement points. Hence, the test should be set up so that there is sufficient flow to cause sufficient head-loss to be measured accurately. There are two different approaches to this:

1. *Monitoring of ambient flow conditions*: If the diurnal flows are sufficient to cause significant head-loss then the field test can be based on the measurement of

concurrent flow and pressure data. This is the common practice in the UK where diurnal velocities can be relatively high and the field test is used to collect not only data for calibration, but also the flow data required to allocate nodal demands. Flows need to be recorded at all sources into the system and at internal zone flow meters. The inlets and outlets of all storage and the usage of the larger metered consumers should also be recorded. Pressures are measured simultaneously typically at between 20 and 35 % of the number of demand nodes.

2. *Monitoring of artificially stressed flow conditions*: If the typical peak diurnal flows are well below the hydraulic capacity of the pipe, then there may be insufficient velocity in the pipe to cause significant head-loss under normal diurnal conditions. To overcome this problem, fire hydrant flow tests can be conducted. Pressure is recorded at appropriate locations upstream of one or more hydrants which are opened and their discharge flow rate recorded. Based on the difference in head-loss between monitoring points as the discharged flow is increased, it is possible to calculate the roughness between monitoring points. The benefit of this method is that it can test greater extremes in hydraulic conditions. The limitations of this method are that:
    - It does not record the diurnal pressure fluctuations which can give an insight into the behavior of the network
    - It will not indicate anomalies with the ambient flow patterns and so will not validate the assumed demand allocation
    - Many tests are required to cover the whole network comprehensively. For this reason the modeler may try to isolate a set of characteristic pipes in this way and assume that each category of pipe has consistent properties throughout the network.

The quality of a calibrated model is directly related to the quality of the data derived from the field test and the number of reference points monitored. Increasing the number of reference points will also help identify anomalies and can reduce the calibration effort. Confidence in the calibration can be improved in a number of ways:

1. There is merit in performing a combination of the two above monitoring methods, especially when building models for water quality simulation where it is important to test the models performance across the full diurnal flow cycle, not just the performance under peak flows
2. All calibration adjustments should be clearly documented and any significant anomalies investigated and rectified or simulated in the model. Anomalies are often caused by major leaks, closed valves, or unknown connections between pipes.
3. Prior to the calibration, the model should be set up to reflect the supply and demand conditions which are contemporary with the calibration data. This will require different demand profiles and flows to those used in a final model which would normally represent average day or peak day flow conditions.
4. In addition to matching pipe flows and pressures, the model should replicate storage tank, pump station, and valve performance. The calibration should test the overall system performance. It is preferable to establish pump and valve performance curves from in-situ testing rather than to rely on manufacturers' data.
5. Tracer studies can be used to determine flow paths and flow rates

6. The diurnal flow patterns can show considerable variability from day to day. It is normal for weekends to have different demand patterns compared to midweek, but smaller variations may occur within the week. Monitoring flow and pressure over several days or, ideally, for a full week allows the modeler to assess the consistency of the network behavior and ensure that representative data is selected for use in the calibration.
7. A consistent method should be used to assign nodal demands or diurnal demand profiles. Altering them as part of the calibration process is inappropriate unless the adjustments can be justified by logical reasoning and supported by field data.
8. In order to limit the number of unknowns before the calibration field test and thereby avoid an abortive field test, extensive preliminary field investigations are advisable. This is particularly important where flow data is recorded for demand allocation purposes when it is necessary to ensure that all flow meters into and out of the zone are operating accurately and all metering zone valves are fully shut. This preliminary work would include flow mass balance checks, zone valve status checks, and installing additional flow and pressure monitoring points.
9. Following calibration, the model should ideally be verified against a separate data set to that which was used to calibrate the model which represents different hydraulic conditions. However, this verification is often omitted due to the additional time and expense involved.

Model Limitations

Hydraulic models have a number of limitations with respect to their flow prediction capabilities. Since water quality models rely on the flow predictions of hydraulic models to give the travel time, these limitations can have significant implications for age of water and water quality modeling:

1. It is normal practice to calibrate models by adjusting the roughness coefficients. However, high head-loss in a pipe could be caused by a reduced cross-sectional flow area due to corrosion or deposited sediment. If the diameters in a model are over-sized, then the velocities will be under-predicted.
2. Sometimes models are simplified by omitting small diameter mains. Where a large diameter main is connected in parallel with a small diameter main which has been omitted from the calculated model, this will result in the predicted flow being greater than the actual flow.
3. It is common to calibrate models against pressure rather than flow measurements (this is the normal practice in the UK). However, where the velocity is low, the pressure will be relatively insensitive to changes in the velocity, reducing the confidence in the predicted velocity. Low flow conditions with their corresponding long travel times are of great significance to water quality parameters.
4. In networks with a dendritic (tree like) structure, the modeled flows at any location are dictated by the modeled demands downstream of that point. However, in looped networks, the flow routing is dictated by a balancing of head losses within each loop. If the length of a loop is short, the head loss along it will be low and so there is reduced confidence in the predicted flow routing.

5. In general, demands are not measured directly. It is commonly assumed that all the domestic properties within each sub zone (hydraulically discrete supply area) have identical demand patterns. In each sub-zone there may be 3000 or more properties and, in practice, there will be significant variability in the demand patterns and quantity of water consumed.
6. Most demands are either on or off (square-wave). It is only when individual consumer demands from a significant number of properties are aggregated that they tend to result in a smooth overall demand profile that may be relatively consistent from day to day. Towards the periphery of the network, the number of downstream properties decreases and hence the actual flows become more erratic and less predictable. Hence, network models are likely to have lower accuracy in the velocities predicted towards the limits of the system.
7. Time varying network models are based on sequential steady state analyses typically at 1-hour or 15-minute time intervals. The flows predicted by the model are therefore average flows across each time step. In practice the actual peaks in flow may be of shorter duration. This is particularly the case towards the limits of the system for the reasons cited in Point 6 above.

Application of Network Models to Water Quality Modeling

Until recently, the success of water quality modeling was restricted by the lack of suitable cost effective monitoring equipment with which to obtain accurate water quality data for model calibration. Studies either had to rely on labor intensive manual sampling or used customized process monitoring equipment which was large, expensive, and lacked the robustness necessary for portable applications. The advent of mobile monitoring equipment designed specifically for installation within distribution networks greatly increases the viability of water quality modeling.

The capabilities of different modeling software vary slightly, but most modern packages can be used to analyze:

- The age of water and hence the retention time within the network
- The amount of water from a specified source (source contribution)
- The blending of conservative substances, the concentration of which does not decay or grow with time
- The time dependant decay or growth of substances
- Propagation of a contaminant injected at any point within the network, for example to investigate a pollution incident.

In addition to the above, some software incorporate models for bacteriological growth and sediment transportation, the limitations of which are discussed later.

Water quality modeling should ideally be based on detailed (all mains) hydraulic models in order to predict velocities and travel times with confidence. Where the hydraulic model has been simplified by omitting small diameter mains, overestimated velocities in the simplified parts of the model will cause the age of water to be underestimated at any point downstream of the simplification. The ability to generate uncalibrated models automatically from GIS gives a cost-effective method of producing all mains models reducing the benefit of simplified models (Twort, Ratnayaka, and Brandt 2000).

CFD Modeling

CFD is generally taken to mean the numerical modeling of the motion of fluids based on the application of basic physical principles. Conventional modeling techniques such as the models of pipe flow incorporated into network models only predict the motion in one dimension i.e. axially to the pipe. CFD can model the behavior of fluid in 3 dimensions. This enables the modeling of eddies and mixing patterns which are of great significance to water quality.

CFD generally involves defining a three dimensional mesh which splits the fluid into a large number of small elements. The behavior of these elements is then predicted using the Navier-Stokes set of simultaneous differential equations describing fluid flow dating from 1845. There are five equations covering:

- Conservation of mass
- Conservation of momentum (three equations, one for each dimension)
- Conservation of energy

In addition to the above, CFD analysis can also model chemical reactions, heat transfer, multi-phase flow (water, gases, and particles), and their interaction, all of which are desirable for detailed water quality modeling.

There are numerous commercially available CFD modeling packages. The software was originally developed for the aerospace and nuclear industries, but is now suitable for widespread application. Hence, most packages are general purpose and not specific to the water industry. CFD is not currently a pragmatic method of modeling an entire distribution network. The computing power required would be excessive. Furthermore, the predictive capability would be no better than that of conventional network models due to the inherent uncertainty in pipe roughness and other unmeasured network properties. The primary application for CFD with respect to water distribution is to model the behavior of storage tanks. In particular they can be used to analyze flow paths through tanks and thereby identify zones with inadequate turnover of water.

Hydrotank

Hydrotank was developed as part of a previous AwwaRF project (Grayman et al. 2000) and the software was included with the report. The software is designed for a simple geometry tank, with an inlet and an outlet. The software simulates both steady and transient flow and can predict disinfectant decay. It is straightforward to use and is a lower cost option than using more sophisticated CFD packages. However, in comparison with such packages, the user is restricted in the complexity of the geometry which can be analyzed and the visualization tools available (Figure 3.2). Internal structures such as columns or baffles cannot be modeled with Hydrotank.

Detailed CFD Modeling Packages

Most commercially available CFD modeling software enables the modeling of more complex geometries than is possible using Hydrotank. They also give more detailed and flexible methods of viewing the results for example; three-dimensional plots of flow paths (Figure 3.3).

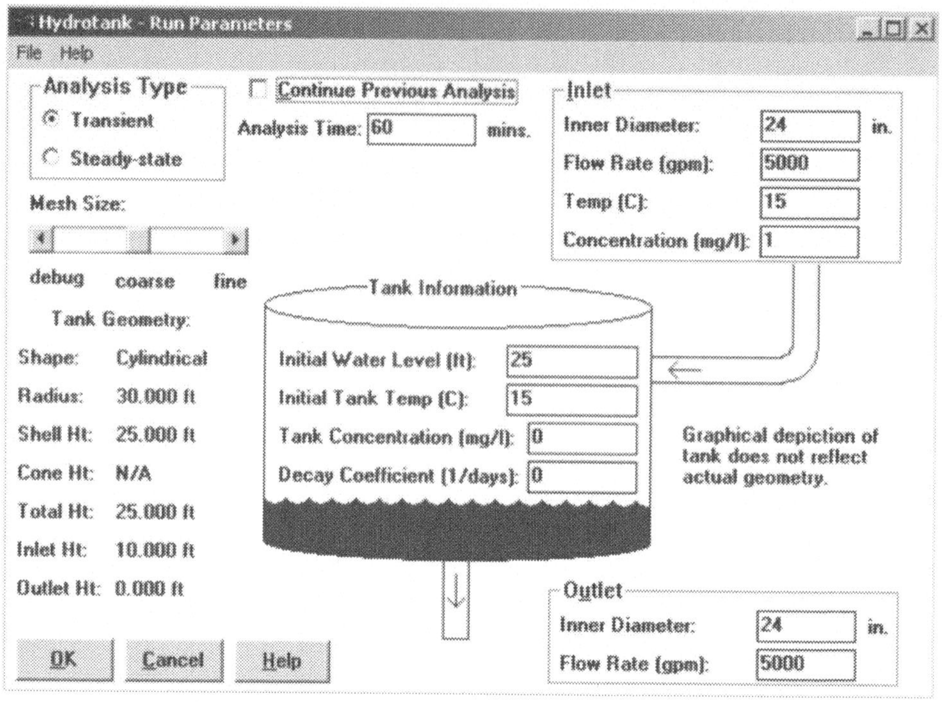

Figure 3.2 Hydrotank graphical display

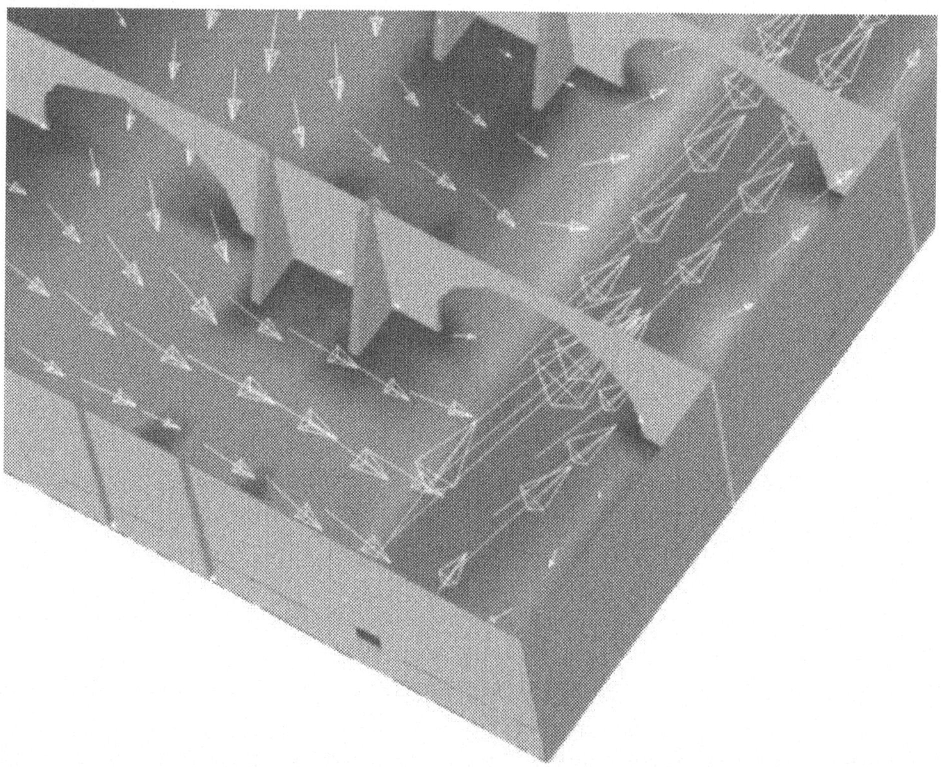

Figure 3.3 Example of CFD modeling of a complex reservoir structure

Commercial CFD packages offer the only effective numerical method of fully modeling the flow patterns within tanks containing internal piers baffles or other structures. The typical steps of using one of these packages to model a storage facility would be:

1. Define the geometry of the structure and generate a computational mesh for the analysis. In many packages this can be done automatically by importing computer aided design (CAD) files.
2. Define the boundary conditions for the model
3. Run the CFD solver to perform the analysis. Even with modern computers this may take several hours.
4. Display the results and review the flow paths. The location of zones in which there is poor mixing and slow turnover of water. It also helps engineers visualize the flow behavior giving them a better idea of any remedial designs to resolve problems.
5. Plot a retention time distribution curve (volume fraction versus retention time)
6. Repeat the exercise if appropriate to analyze alternative modes of operation or the effect of adjustments to the design; for example, the installation of baffles

Front ends are available for some of the detailed CFD packages which aim to simplify the modeling of storage facilities. These have some similarities with Hydrotank. They have the advantage that people without extensive experience of CFD will be able to use them, but because it is still necessary to purchase the full CFD package, they are likely to be expensive with respect to their capabilities. One such application is TankSim which is a customized front end to the CFD software produced by CFX (2001). The software is operated from a graphical user interface, generating results that are suitable for viewing with any standard web browser. The software predicts mixing performance by running a dye tracer test. TankSim also models transportation and deposition of sediment, chlorine concentrations, and water age.

Benefits and Limitations of CFD Modeling

The benefits of CFD modeling include:

- It is the only pragmatic method of analyzing flow paths within complex structures.
- It is suitable for the analysis of common inlet and outlet or standpipe tank designs and where there is poor mixing within the tank.
- The software enables the modeling of complex interactions between water quality parameters, providing the behavior can be defined by numerical rules.
- Many of the more detailed software packages have sophisticated graphical methods for viewing results which allow users to gain a better understanding of the three dimensional behavior within the tank.
- Since the models are primarily based on fundamental physical principles, they are more reliable than more empirical models if the boundary conditions are significantly altered.
- The models allow dynamic inertial effects to be predicted which can be significant in tanks. If a tank has been standing for some time with no flow in or out, the flow path upon restarting flows can be significantly different from the flow paths that would occur if it was operated with a steady through flow.

The limitations of CFD modeling include:

- High cost of software
- The need for skilled modelers
- It can be time consuming to model complex geometries
- The above means that CFD modeling tends to be a relatively expensive exercise

Tracer Studies

Tracer studies involve injecting a chemical into the water supply for a fixed period. The water is then monitored at downstream points to determine the duration before water containing the chemical passes the monitoring points. Guidance on the use of tracer studies to assess the hydraulic characteristics of water treatment facilities is given in a previous AwwaRF report (Teefy 1996) and similar techniques can be used for distribution storage facilities. When performing tracer studies on tanks, the profile of the concentration of the tracer chemical against time indicates the mixing characteristics. Tracer studies can also be used to measure retention times within distribution networks.

Accurate information about the flow patterns within distribution networks is essential when modeling water quality. Flow meters only give information about a single location and the number of points at which it is viable to install permanent or temporary flow metering within a network is limited by cost. Pressure monitoring can only give accurate information about flow patterns where the hydraulic gradient is high. Tracer studies offer the flexibility to determine the travel time to virtually any point within the network. The use of tracers as a means of increasing the accuracy of a hydraulic model, especially when the hydraulic model is to be used for water quality modeling, is documented in the literature by Clark and Grayman (1998), Grayman (2001), Cesario et al. (1996), and Powell et al. (1999).

The trace chemical must be non-toxic and approved for use in drinking water. Utilities will often prefer to use chemicals which are routinely added to drinking water supplies. The trace chemical should be easy to measure accurately at low concentrations using a portable monitor which logs the recorded data. If the readings can be taken in the field once the trace chemical has passed the monitoring point, the monitor can be relocated further downstream. In this way the trace can be captured at several different locations using one piece of monitoring equipment. In reticulated systems, a trace profile will be obtained for each of the flow paths. In such circumstances it is beneficial to measure and log the trace concentration at a relatively high frequency so as to improve the resolution of the trace chemical.

The following trace chemicals have been widely used in distribution systems:

- Fluoride: Where fluoridation is practiced a trace can be achieved by turning off the fluoridation for a short period.
- Chlorine: A trace can be easily inserted to any chlorinated supply by inducing a short (15 minute) small (0.2 mg/l) increase or decrease in the dose. Chlorine can be readily measured using widely available equipment. The major disadvantage is that the concentration decays, so it can only be used to give retention times in regions which maintain some residual chlorine.
- Sodium chloride: Sodium chloride directly affects the conductivity of water which is easy to measure using relatively cheap equipment.

It is necessary to take account of local conditions and regulations, which may prohibit the use of certain trace chemicals. It is also important to consider whether the trace chemical risks an adverse effect on the distribution system for example the destabilization of corrosion deposits.

A study by Benrahnem (1996, cited in Desjardins et al. 2000) found that tracer results can still lack precision owing to a number of distribution features including: the number of possible paths feeding the sample location, mixing of the tracer coming from the various paths, dispersion or diffusion of the tracer, and the non-steady flow of the distribution system.

Skipworth, Machell, and Saul, (2002) developed an automated and easily repeatable procedure for determining the empirical travel time in a distribution network based on the injection of a tracer. Sodium chloride was introduced into a network for a short period at a set concentration when the flow rate was reasonably stable. The concentration of the sodium chloride was then determined at ten downstream locations using conductivity meters and the travel time estimated. The authors then compared the modeled age of water with the empirically determined age using the tracer. This comparison highlighted inaccuracies in the model where the flow regime in the network was complex.

Recent studies with tracers have also investigated the concentrations of the tracers and employing more than one tracer (Creasy, Chambers, and Russell 1997). Desjardins et al. (2000) injected tracers in pulses into the distribution network so that the pulse frequencies can be identified, thereby being able to improve the accuracy of calculating the retention time.

Several studies have successfully used the ambient variation in water quality parameters (turbidity and chlorine) within a water as the trace signal. The resolution of the trace may be less clear, but the method avoids any concerns about injecting a substance into the system.

There is potential to induce a regular trace signal into a system which is recorded within the network using permanently installed equipment. Software could be written to automatically recognize the trace enabling on-line measurement of retention time.

Despite the value of tracers in validating hydraulic models and water age calculations, routine application in distribution networks is relatively rare. A number of reasons have been suggested:

- Lack of automated monitoring devices for some tracers
- High cost and effort associated with the manual sampling required
- Approval required from industry regulators to introduce tracers (fluoride, sodium chloride, lithium chloride) into water supplies
- Lack of studies on larger distribution systems
- Cost benefits of tracer studies in systems where flows may frequently be changing

Physical Models

Physical models tend to be used to confirm the design of a structure or the conclusions from analytical modeling. In a recent AwwaRF study, Grayman et al. (2000) demonstrated that physical (scale) models provide an effective and relatively inexpensive means of studying mixing behavior in storage tanks. Dyes or tracers can be introduced into the model inflow and mixing patterns observed, photographed, or recorded using specialist monitoring equipment. The models can be constructed from a range of materials with the opportunity to use transparent materials where dyes are used as the tracer.

Physical scale models are still widely used in the scientific community despite advances and improvements in computer-based modeling techniques. For distribution networks physical models have been built mainly to represent storage tanks, reservoirs, and pump sumps. The advantage of a physical model is that it provides observable evidence in contrast to a numeric solution from a numerical model. The disadvantage of physical models is that hydraulic properties do not scale linearly with physical dimensions. Hence, the model may need to be large for it to be reasonably representative.

Introducing a dye into the inlet water allows visual observation of the flow in a model reservoir. The residence time distribution of the outlet water can be obtained by chemical analysis of the tracer.

Physical models are valuable in validating CFD models and providing confidence for running more complex simulations (Hague, Ta, and Biggs 1999). Advanced flow measurements incorporating laser technologies and particle image analysis (particle image velocimetry, PIV) are possible with small-scale models (Figure 3.4). In comparison to the dye test, PIV provides improved resolution.

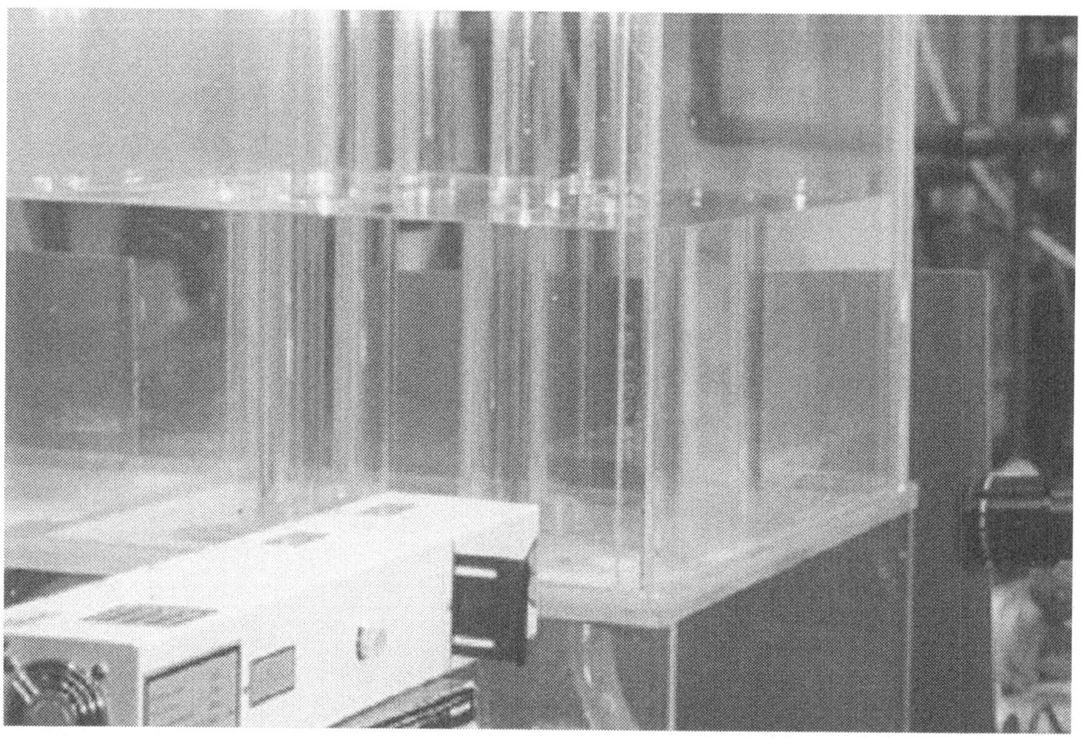

Figure 3.4 PIV measurement on Perspex model of service reservoir

TOOLS FOR ASSESSING WATER QUALITY CHANGES IN RELATION TO RETENTION TIME

These tools allow the user to assess whether a given retention time is liable to result in acceptable water quality.

Laboratory Systems

A number of techniques have been developed for simulating in a laboratory, the water quality changes which occur within distribution systems. The simplest tool developed is the SDS test for THM formation (APHA 1994) (Figure 3.5). With this procedure, THM formation potential is estimated in the laboratory by incubating water in a glass vial. This approach does not take into consideration the effects of the distribution system hydraulic regime and pipe wall interactions but determines changes in bulk water properties.

The WRc utilized SDS systems to quantify the "shelf-life" of treated water in terms of DBPs and biological growth (WRc 1997a). The purpose of the study was to determine changes in microbiological and chemical quality of drinking waters in the absence of any influence from the distribution system. The variables studied were DBPs (THM and HAAs), biological growth, disinfectant residual (chlorine and monochloramine), and taste and odor. These were studied in relation to temperature, water type, biofilms, and time.

Two identical volumes of water were collected after final disinfection from 12 locations. The disinfectant residual in one volume was neutralized immediately and this comprised the initial (day zero) sample. The other volume was stored in bottles for 7 days at 20 °C without neutralization, and this comprised the stored (day seven) sample. The four waters that exhibited most change in quality were selected for more detailed studies based on similar experimentation. Samples from the four water sources were stored in bottles under controlled conditions for two weeks. The microbiological quality, the concentration of DBPs, and the taste and odor were periodically monitored. Experiments examining the rate of chlorine decay and the effect of biofilm were also conducted. The main results of the study were as follows:

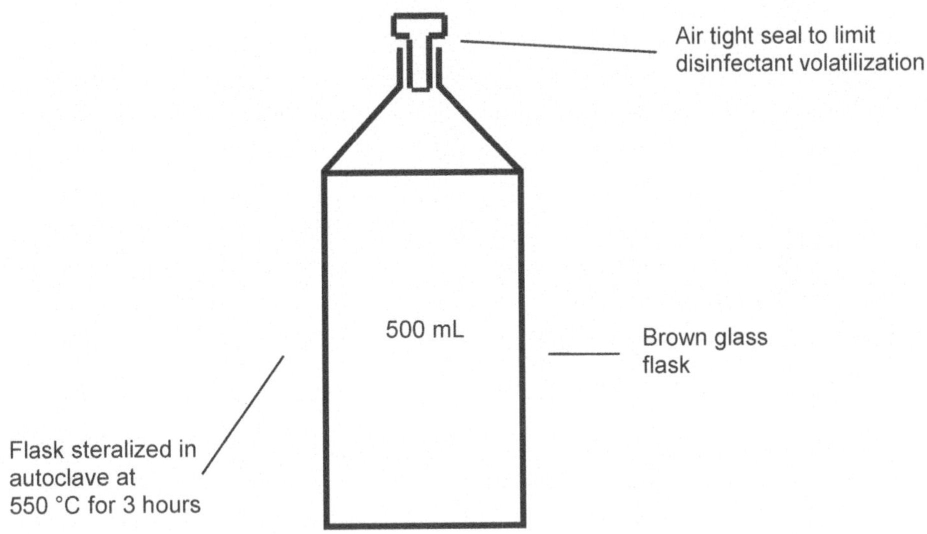

Figure 3.5 Schematic representation of the flask test

Microbiological changes: Using standard monitoring procedures, no microbiological deterioration was detected in waters stored for at least 14 days at 20 °C even though the free chlorine residual declined to less than 0.1 mg/L. The presence of 0.1 to 0.2 mg/L total available chlorine during most of the storage period contributed to the prevention of bacterial growth. Using a more "sensitive" Reasoner and Geldgar agar (R2A) culture method, bacteria were found to grow in some water samples stored for 7 days at 20 °C.

The presence of biofilm gave increased numbers of bacteria in water particularly when the chlorine residual was neutralized at the start of the storage period. In the presence of biofilm, the increased numbers of bacteria detected in water samples was caused by the removal of attached microorganisms rather than growth in the water phase.

Formation of THMs: The extent of THM formation was dependent on the TOC present in the source water, the treatment applied at the treatment works, and the product of the chlorine concentration and the contact time (CT). The rate of THM formation increased with chlorine concentration and temperature. Increases in the concentration of THMs were proportional to the additional CT value arising from storage. The presence of a biofilm had no significant effect on the concentration of THMs formed during storage.

Other DBPs: For the majority of the waters studied the concentrations of aldehydes, trichloroacetic acid, and dichloroacetic acid did not increase to significant concentrations during seven days storage.

Chlorine residuals: The rate of decay of chlorine residuals was very dependent on temperature and chlorine concentration. In stored water the decay of chlorine residual did not follow first order kinetics.

Taste and odor: Factors which appeared to affect the occurrence of taste and odors are an increase in the numbers of heterotrophic bacteria to above 10^3 per mL (7-day incubation on yeast extract agar) and the duration of storage.

Nutrients: Large decreases in the concentration of nutrients (AOC and BDOC) were found to occur when no bacteria (as measured by standard culture techniques) were present in the water. The extent of this non-biological removal is independent of temperature.

From these findings the authors of the report produced guidelines to assist operators in controlling these water quality problems in distribution networks.

Many more laboratory scale systems have been described in the literature:

- Lab-tubular systems, simple chemostats with submerged blades, and jar tests (Lund and Ormerod 1995, Mackerness, Colbourne, and Keevil 1991, Mackerness et al. 1992)
- Closed loop systems (Lee, O'Connor, and Banerji 1980, McCoy and Costerton 1982, Norman et al. 1977, Pedersen 1982)
- Rototorque (Characklis 1990)
- Annular reactors (Camper 1996)
- Propella (NanCIE 1994)

The system that has found general acceptance is the annular reactor shown in Figure 3.6. While its value for determining kinetics has been questioned, it continues to be used as a tool for assessing the impact of various parameters affecting biological growth and corrosion (Camper 1996; Parent et al. 1996).

Figure 3.6 Annular reactor (*Source:* BioSurface Technologies Corp.)

In the mid 1990's, NanCIE developed a reactor, using a propeller-type centre piece inside of a distribution pipe (NanCIE 1994). The advantage of this system is that the hydraulic residence time is kept independent of the shear stress when operated within a range of rotational speeds. Advantages are the size, the flexibility (ability to test different types of material, both new and used, temperature control, a large array of possible shear stress, and residence time combinations, etc.), and the relatively low cost.

The relative contribution to disinfection demand of the water and at pipe surfaces (material, biofilm, deposits) was determined in a series of experiments using the propeller type reactor (NanCIE 1998). The studies allowed classification of materials into the following order according to their chlorine and monochloramine demand as determined by kinetic constant, from 0.001 h^{-1} to 30 h^{-1} (0.024 d^{-1} to 720 d^{-1}):

- Polyethylene (PE) (0.001 h^{-1} or 0.024 d^{-1})
- Polyvinyl chloride (PVC)
- Epoxy lined ductile iron
- Cement lined ductile iron
- Bitumen lined ductile iron
- Grey iron and ductile iron (30 h^{-1} or 720 d^{-1})

In all situations tested, chloramine was four to ten times less reactive than chlorine. Corrosion by-products and corroded surfaces were found to be highly reactive with chlorine and chloramine, reinforcing the need for corrosion control in drinking water distribution systems.

The study provides direct information on potential oxidant consumption of pipe walls and a standard procedure for future testing of new and old pipe materials.

Pilot Distribution Systems

There has been extensive research into the effects of distribution system conditions on water quality using laboratory pipe rigs. These rigs have gradually increased in complexity culminating in small flow through pipe based systems (Le Chevallier, Lowry, and Lee 1990 and Le Chevallier et al. 1993). The first system attempting to simulate distribution mains conditions where there is a continuous water supply on a relatively large pilot scale was designed and built by NanCIE in 1986 (Colin et al. 1987). This comprises a combination of 6 loops connected in series where the entire system simulates plug flow conditions similar to a real distribution system. The individual loops provide mixed reactor conditions representing points in the network with increasing residence time as the water progresses from loops 1 to 6. There has been an extensive program of research with the NanCIE pilot rig and this has led to an increased understanding of the factors influencing microbiological water quality in distribution (Mathieu et al., 1992; Block et al., 1993). Other industrial size pilot systems have been built, including continuous-feed, looped systems (AwwaRF and DVGW 1985, Camper et al. 1993, Piriou 1994, Camper 1996, Colin et al. 1987, Jackson and Sheiham 1980, Lee, O'Connor, and Barnerji 1980, Piriou and Lévi 1994, Prakash et al. 1988), and open pipe rigs (Holt et al. 1994, LeChevallier 1990).

Holt et al. (1994) designed and built a pipe rig specifically for studying the effects of microbiological factors on water quality in distribution systems. The 0.8 mile (1.3 km) long, single pass pipeline is the largest pilot-scale system currently available for the fundamental study of distribution systems. The pipe rig primarily consists of 110 mm (4 in.) internal diameter MDPE pipe, buried between 0.5 and 1.0 m (1.6 to 3.3 ft.) underground in three sections of 400 to 500 m (1312 to 1640 ft.) in length (Figure 3.7). Between each buried section, water passes through 24 m (78.7 ft.) of pipe situated above ground, providing an opportunity to monitor water quality at different retention times. The three above ground sections are located in a building and consist of different 110 mm (4 in.) diameter pipe materials i.e. ductile iron, cast iron and MDPE (Figure 3.8). Each section has on-line water quality monitors measuring chlorine, chloramine, flow, pH, redox, and temperature. Each section also has sampling taps and bioprobes with coupon inserts, flush with the internal pipe wall, enabling spot sampling of overlying water quality and surface features respectively. The 1 cm^2 coupons can be made of different materials, notably iron and plastics.

The pipe rig simulates the conditions that drinking water experiences when traveling through low flowing regions of a distribution network. It has been used extensively to study biofilms on different materials (Delanoue et al. 1997), and under different hydraulic regimes (McMath et al. 1999a, Maier et al. 1999), nitrite formation (Holt et al. 1995), coliforms (McMath et al. 1999a and 1999b), particle entrainment (Woodward et al. 1995, Maier et al. 1999), and red water generation (Smith et al. 1998). It provides a unique environment for controlled experiments at a scale comparable to a small distribution system. Findings from the pipe rig have been used to develop control strategies that have been implemented in full-scale distribution systems (Holt, Smith, and Colbourne 1995, Smith et al. 1998, Williams, Andrews, and Wakeford. 2001).

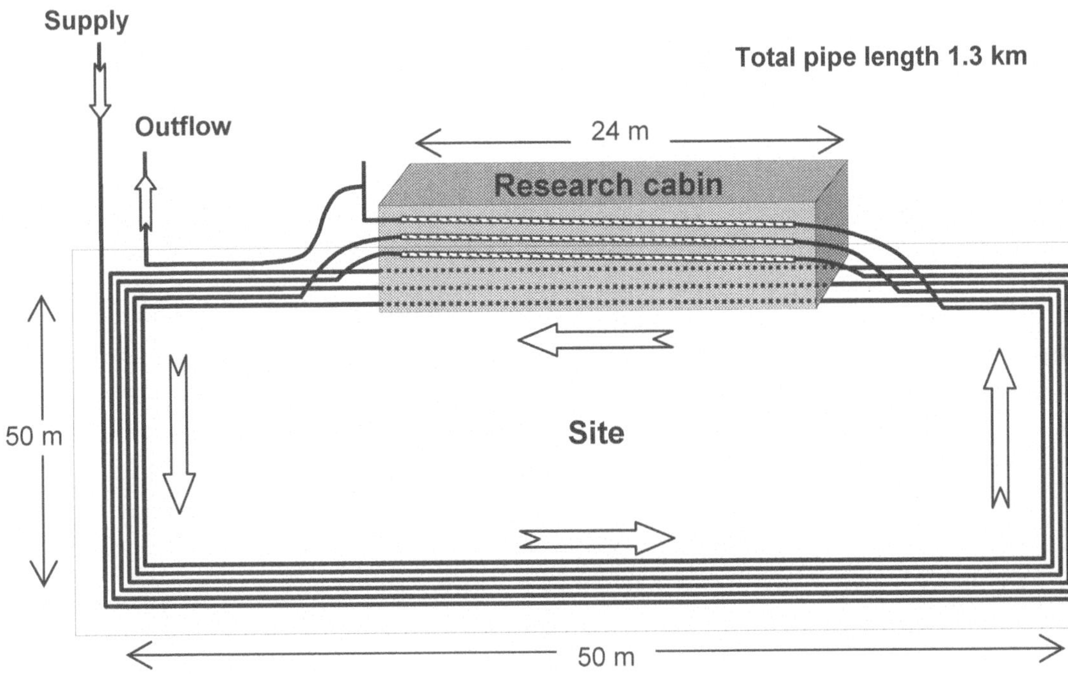

Figure 3.7 Pilot distribution system for studying water quality

Figure 3.8 Experimental station showing instrumentation bioprobe device

Full Scale Studies

If the retention time within a network can be readily determined, then the best method of ascertaining the maximum acceptable retention time may be to investigate the response of that network directly. This would involve collecting water quality data from a number of locations throughout the network and ascertain the age of water corresponding to each sample. It may be possible to base this on historic regulatory samples. Unfortunately previous studies have demonstrated that it is often not simply a case of plotting water quality versus retention time and establishing a correlation. In a network of similar pipes, and with parameters which progressively deteriorate such as free chlorine residual, this may work, but more frequently it will produce a cloud of data with no clear trend. The reasons for this are:

- Microbial water quality will not necessarily deteriorate with retention time, rather it is the risk to water quality that will increase with water age.
- Water quality is dependant on the integrated effect of the retention time in the different pipe types throughout the system. Retention times should only be directly compared where the physical characteristics are similar.

These problems can be addressed by:

- Characterizing the predominant pipe properties upstream of each sample and only comparing data for water which has resided in similar pipes
- Identifying the lowest retention time corresponding to an unsatisfactory water quality

TOOLS FOR PREDICTING WATER QUALITY BASED ON RETENTION TIME

This Section describes those tools that first calculate retention time and then utilize algorithms to predict water quality changes. In theory, any tool for estimating retention time could be combined with algorithms that describe age related water quality changes. Water quality algorithms have been incorporated into a number of hydraulic modeling packages to provide the additional functionality and many different 'water quality models' have been developed ranging in complexity.

Applying these models to full distribution networks may be complex and time consuming for a number of reasons:

- The models need to incorporate details of all the different elements in the network. Simplification which affects the hydraulic performance of the model will distort the water quality analysis. Therefore all main (100 % detail) models should be used.
- The model parameters required for each element in the network may not be readily available.
- Models which have been produced as the result of detailed research studies may not have been developed into a user friendly software package for practical application.

Consequently; the confidence in an analysis may not justify the modeling effort required to build a model of a complete distribution system.

A more pragmatic approach could be to use the basic models in reverse and calculate the retention time corresponding to a minimum acceptable water quality for a range of characteristic model parameters. The analyst could then identify the maximum acceptable age of the water for the conditions in the network and run a hydraulic model to identify any regions in which this age is exceeded.

Such models might be a simple correlation derived from laboratory, pipe rig or full scale trials undertaken specifically to investigate the problem of concern. Alternatively the model may be taken from published research.

Batch Models

Predicting water quality changes within a distribution network are complicated by the need to account for the propagation of the water through the network and the non-uniform properties of the pipes, reservoirs, and other distribution assets within it. The majority of water quality models simply predict the changes that occur in a single vessel of water as this is much simpler to model and analyze. Such models are referred to herein as 'batch models'. Some models will also account for water quality changes that occur due to interactions with the walls of the vessel, but most assume the vessel to be inert and so only model the behavior of the bulk water. Batch models have a number of applications:

- Assessing the maximum acceptable retention time for a given set of network conditions (See previous sections of this chapter)
- Predicting water quality within reservoirs, although only if the reservoirs are completely mixed which is often not the case
- If the water quality parameter is controlled by reactions in the bulk flow rather than at the pipe wall, the network properties may be insignificant, in which case these models may still give reasonable predictions of water quality within distribution
- If the pipe properties are relatively consistent, it may be possible to model pipe wall reactions by using surrogate or modified values for bulk water parameters
- These models form the basis of more complex network water quality models (see subsequent sections of this chapter)

Reaction Kinetics

The rate of most chemical reactions decreases as the concentration of the reactants decreases and is typically defined by an equation of the following form:

$$\frac{dC}{dt} = -k\,\mathrm{f}\{Reactants\} \tag{3.6}$$

where C = concentration
k = reaction rate constant (units dependent on order of reaction)
$\mathrm{f}\{Reactants\}$ = some function of the concentration of the reactants

The reaction order is defined as the order of the differential rate in Equation 3.6. Where the reaction rate is directly proportional to the concentration of a single reactant, the reaction is classed as 'first order'. A 'second order' reaction can be due to the reaction rate being proportional to the square of the concentration of a single reactant, or due to the reaction rate being proportional to the product of two reactants.

Many chemical reactions are complex, involving more than two different compounds and numerous intermediate steps. This means that the overall reaction kinetics will be complex and difficult to define. For pragmatic reasons therefore, it is often necessary to simplify the reaction kinetics. This can be justified where one part of the reaction is much slower than the rest so that it becomes the 'rate determining step'. Alternatively, where all but one of the reactants are present in large excess, and only the concentration of that reactant changes significantly during the course of the reaction, the reaction rate may appear to follow a first order pattern. This is sometimes referred to as a 'pseudo first order reaction'. For example, free chlorine decay is frequently modeled by a first order kinetic model (exponential decay with time):

$$\frac{dC}{dt} = -kC \tag{3.7}$$

or $$C_t = C_0 \exp(-kt) \tag{3.8}$$

where C_t = concentration at time t
C_0 = initial concentration

Bulk and Wall Reactions

It is widely accepted that the concentration of many water quality parameters will vary due to reactions with compounds contained within the bulk water and due to reactions at the pipe wall. The overall reaction coefficient is defined as the sum of a bulk and wall coefficient.

$$k = k_b + k_{w*} \tag{3.9}$$

where k_b = bulk water decay coefficient
k_{w*} = wall decay coefficient

It has been argued that the rate of consumption at the pipe wall will be controlled by the reactivity of the pipe surface and the rate at which the reactant can be transferred from the bulk flow to the pipe wall. This is accounted for in the widely used EPANET model (Equation 3.10), in which the rate of radial mass transfer is defined by the coefficient k_f which is itself a function of turbulence and the molecular diffusivity.

$$k_{w*} = \frac{k_w k_f}{R_h (k_w + k_f)} \tag{3.10}$$

where k_f = mass transfer coefficient
k_w = wall reactivity coefficient

Free Chlorine

Chlorine decay is frequently modeled as a first order reaction and this has often given reasonable results. A variety of alternative models for bulk decay were tested in the AwwaRF study by Vasconcelos et al. (1996) into chlorine decay. Some of these were shown to give better fits against observed data than the first order model. However, the better fit was often a consequence of the models containing more adjustable input variables and thereby greater mathematical degrees of freedom. Multi variable models are of limited benefit unless the coefficients can be related to measurable parameters. Similar tests by Powell et al. (2000a) on over 200 observations of bulk decay also found multi parameter models to give a better fit, than the first order model. Nevertheless the first order model generally gave a good fit with a mean R^2 correlation coefficient of 0.95.

There are, however, some limitations in using a first order model (Powell et al. 2000b):

1. It tends to underestimate the initial chlorine decay in the first one to four hours following chlorination
2. First order decay coefficients have been observed to decrease if the initial chlorine dose is increased or if the water is re-chlorinated
3. The rate of decay has been shown to relate to the organic content of the water. This is demonstrated by the observation that the rate of bulk decay when distilled water is dosed with chlorine is virtually negligible.

The above limitations can to some extent be accounted for by such measures as:

- Considering the decay in two phases ('short term' and 'long term' demand)
- Defining the decay coefficient as inversely proportional to the chlorine dose
- Defining the decay coefficient as linearly proportional to TOC or UV

The primary cause of the above limitations is that the first order model fails to account for the concentration of the compounds with which the chlorine reacts and as these compounds react and diminish in concentration, the rate of reaction will be reduced. This has lead a number of researchers to promote the use of a second order model which accounts for the concentration of chlorine and a second reactant (Jadas-Hécart et al. 1992, Hua et al. 1999, Clark and Sivaganesan 2002). Furthermore, because the concentration of the second reactant is likely to be related to the concentration of TTHM pre-cursors, it is possible to characterize TTHM formation as a function of chlorine demand (Clark 1998). A second order model involves additional model coefficients associated with the concentration and reactivity of the second reactant. However, Clark and Sivaganesan (1998) have developed models which relate these additional coefficients to measurable parameters such as TOC, pH, and temperature. This increases the viability of second order models and may lead to their increasing use.

Chloramine

Like free chlorine, chloramine will progressively decay with time, but the rate of decay is much slower. Hence, chloramine is often modeled in a similar manner to free chlorine, but using lower decay coefficients. NanCIE (1998) showed that the decay due to reactions with pipe materials were lower for chloramine than for free chlorine by a factor of up to ten.

Some of the most extensive research into chloramine decay has been that undertaken by Valentine, Ozekin, and Vikesland (1998). They found that it was possible to model the decay using a mechanistic model of fourteen interrelated chemical reactions. However, they also concluded that it was generally adequate to use a much simplified second order model.

$$\frac{1}{[NH_2Cl]} - \frac{1}{[NH_2Cl]_0} = k_{VCSC}t \qquad (3.11)$$

where $[NH_2Cl]$ = monochloramine concentration at time t
$[NH_2Cl]_0$ = initial monochloramine concentration
k_{VCSC} = Valentine chloramine stability index

The value of k_{VCSC} varies with pH, initial chloramine concentration, inorganic carbon, and temperature. However, it can be calculated from measurable parameters and known rate constants and so does not need to be determined by experiment for each water.

DBPs

Most modeling of DBPs has focused on total THM formation. Although some of these models have also been applied to HAA formation, only one model reviewed in the following section was specifically developed for HAAs (Westerhoff et al. 2002, mechanistic model). This model simulates THM and HAA formation within treatment plant. All the following models assume DBPs to grow within distribution. However, HAAs have, under certain conditions, been observed to decay, possibly due to pipe wall processes.

THMs formation reactions can occur rapidly and may be substantially complete prior to entering the distribution system in which case they may be treated as a conservative substance. More commonly there will be some continued reaction within distribution with the result that THM growth within distribution can be as significant as their growth within treatment plant. Clark et al. (1994) demonstrated that THM growth tended to decrease with time asymptotic to a maximum formation potential. This can be modeled using a first order saturation growth model (Equation 3.12).

$$\frac{dC}{dt} = k_b(C_L - C) \qquad (3.12)$$

where C_L = limiting concentration

The limitation of this model is that it does not take account of the concentration of the various pre-cursors which control THM formation. Hence the growth coefficient will be highly site specific and if the water is not of a consistent quality the coefficient may show considerable variability over time. To some extent this limitation is overcome by the method proposed by Clark and Sivaganesan (1998). This simulates chlorine decay and the formation of by-products as second order model and assumes that the by-products represent total THMs. Expressions were developed for estimating some of the parameters in the second order model using TOC, pH, and temperature measurements.

More detailed models have been developed for the formation of THMs within the water treatment process, and in particular during super-chlorination. Amy et al. (1987) proposed an empirical model of the following form which has subsequently been incorporated into the USEPA water treatment plant (WTP) model:

$$[THM] = X_0[TOC]^{X_1}\,pH^{X_2}t^{X_3}T^{X_4}[Br]^{X_5}[Cl_2]^{X6}[UV]^{X_7} \tag{3.13}$$

where $[THM]$ = THM concentration at time t
X_0 to X_7 = WTP model fitted coefficients
$[TOC]$ = TOC concentration
T = temperature (°C)
$[Br]$ = bromine concentration
$[Cl]$ = chlorine dose
$[UV]$ = UV concentration

The value of the coefficients varies depending on the THM of concern and will to a certain extent be site specific. Table 3.1 lists the values for the coefficients fitted by Westerhoff et al (2002) based on a series of approximately 100 chlorination experiments. The model has also been applied to both THM and HAA formation.

The model was derived from data collected from the treatment process rather than distribution where there were higher chlorine doses and shorter contact times and hence their validity to the conditions within distribution is uncertain. Nevertheless, models of this type have been applied to distribution with some success for both THMs and HAAs (Seidel et al. 2002). A difficulty in applying this model to distribution networks is that the concentration of the variables may not be known. In addition, if the model coefficients (X_0 to X_7) are derived from bottle or SDS tests (see Laboratory Systems), they will not account for interactions with the pipe wall or sediment which Valentine and Vikesland (2000) found could be significant.

Table 3.1
Fitted values for coefficients to THM or HAA formation model

Parameter and water source	Coefficients to Equation 3.13*†							
	X_0	X_1	X_2	X_3	X_4	X_5	X_6	X_7
THM_4								
Source 1	11.0	-0.101	1.20	0.250	0.408	0.367	0.309	0.689
Source 2	28.8	-0.518	1.40	0.281	0.191	0.270	0.325	0.827
Source 3	0.997	0.536	1.23	0.242	0.201	0.0365	0.615	0.0031
Source 4†	27.9	-0.328	1.32	0.257	0.197	0.122	0.261	0.858
Di-HAA								
Source 1	231	-0.162	0.124	0.218	0.117	0.0318	0.323	1.14
Source 2	2.73E7	-1.43	-0.961	0.187	-0.301	0.296	-1.73	2.57
Source 3	1.98	1.39	-0.438	0.220	0.0239	0.192	0.438	-0.626
Source 4‡	4.04E5	-0.205	-2.66	0.132	-0.260	0.593	-0.758	0.512

Source: Data from Westerhoff et al (2002)
* When Equation 3.13 used to model Di-HAA, substitute [THM] with [DiHAA] where [DiHAA] = concentration of Di-HAA
† The units of the coefficients vary with their value. They are consistent with the use of the following units in Equation 3.13:

[THM]	(µg/L)	[DiHAA]	(µg/L)
[TOC]	(mg/L)	T	(°C)
[Br]	(mg/L)	[Cl]	(mg/L)
[UV]	(cm-1)		

‡ Source 4 is a composite of sources 1, 2 and 3

Westerhoff et al (2002) have developed a detailed mechanistic model of DBP formation (THMs and HAAs). The model simulates reactions between chlorine, bromine, and NOM, with the NOM subdivided into instantaneous, fast and slow reacting components. The model incorporates over 70 terms, but they have been parameterized through experimental work so that appropriate estimates of their value can be used when applying the model. The model was found to be more accurate than the empirical model developed by Amy et al (1987) (Equation 3.13) when applied to source specific waters, but less accurate when applied to variable water sources. The model is intended for application to treatment plants and the authors are unaware of its application to distribution systems.

Microbial

The earliest models for the inactivation of microorganisms by disinfectants were the Chick and Watson models proposed in 1908 (Equation 3.14) and a generalization made by Hom in 1972 (Equation 3.15) (cited in Haas and Joffe 1994):

$$\ln\left(\frac{N}{N_0}\right) = -k^* C^{n'} N \qquad (3.14)$$

$$\ln\left(\frac{N}{N_0}\right) = -k^* m C^{n'} t^{m-1} N \qquad (3.15)$$

where N_0 number of microorganisms at time zero
N number of surviving microorganisms at time t
C concentration of disinfectant
n' coefficient of dilution
m Hom disinfection kinetic parameter
k^* pseudo first order reaction rate constant

More recently Haas et al. (1996) combined the Chick Watson and Hom models with first order decay model to produce:

Modified Chick-Watson model:

$$\ln\left(\frac{N}{N_0}\right) = \frac{-kC_0^{n'}}{n'k^*}\left[1 - \exp\left(-n'k^*t\right)\right] \qquad (3.16)$$

Modified Hom model:

$$\ln\left(\frac{N}{N_0}\right) = -\left(\frac{m}{n'k^*}\right)^m kC_0^{n'}\left[1 - \exp\left(\frac{-n'k^*t}{m}\right)\right]^m \qquad (3.17)$$

Longer contact time and higher disinfection concentrations cause greater inactivation of microorganisms on contact with chlorine. The Chick Watson model produces an adequate fit of inactivation by free and combined chlorine under demand free conditions according to results of Haas and Karra (1984), however Pernitsky et al. (1995) stated that the Hom model was found to be better in all disinfection streams studied. Ransome et al. (1995) found that the rate of inactivation is not linear and that the response showed an initial rapid kill which diminished after a 90 % reduction in bacterial numbers had been achieved.

Attempts have been made to model microbial variations in distribution systems. These range from highly empirical models to statistical, black box models. A deterministic model was produced by Lu, Biswas, and Clark (1995) incorporating a range of interactions within a distribution system. This two-dimensional transport model is highly complex, accounting for the simultaneous transport of substrates, disinfectants, and microorganisms. The authors concluded that three significant factors were important in influencing water quality. Firstly, shear stress is significant as it determines the rate at which biofilm sloughs from the pipe surface. Secondly, radial diffusion in turbulent flow may either promote biofilm growth by providing nutrients at the pipe surface or inhibit growth by transferring disinfectant. The relative magnitude of both nutrients and disinfectants is therefore significant. Finally, pipe materials with a high chlorine consumption rate will cause low chlorine concentrations at the pipe-water interface and so not control biomass accumulation.

Gatel et al. (1995) developed an empirical model to predict the bacteriological water quality in a given week if the water quality of the previous week is known. The model was based on data collected in the suburbs of Paris, 1992-1994. Results concluded that the model was strongly influenced by the general biomass quantity as measured by epiflorescence. Temperature and chlorine appear to be less important factors, however this partly reflects the events chosen to design the model. The authors also hypothesize that biofilm was acting to protect bacteria from disinfection.

The SANCHO model developed by Servais et al. (1995) predicts fixed and free biomass as measured by epiflorescence microscopy against residence time. It is based on considering a distribution system as a continually stirred tank reactor and has been applied by Laurent et al. (1997) to three distribution systems. The surveys gave a reasonable correlation between modeled and observed concentration of free biomass ($R^2 = 0.69$). However, the validity of the approach has been questioned by Gagnon et al. (1997).

Servais et al. (1999) further developed the SANCHO model to predict the changes in chlorine concentration, BDOC, and the biomass of free and attached bacteria in pipes where the diameter decreases as a function of residence time. The variables in the model include free floating bacteria, adsorbed bacteria, three different fractions of BDOC, and free chlorine. Reaction kinetics and numerical values for the variables were determined from laboratory experiments.

The SANCHO model calculates the changes in BDOC, chlorine concentration, and free and fixed bacteria under steady state conditions in relation to retention time. Mary-Dile et al. (2000) developed an interface between SANCHO and a hydraulic network model. Water residence time information was extracted from the hydraulic network model and used by SANCHO to estimate the bacteriological quality of water at any given location in the network. The simulations showed that hydraulics can account for a large proportion of the variation in bacterial numbers that are reported in the distribution network. The authors also indicated that models for THM formation could be developed based on similar principles.

Storage Tank Water Quality Models

If a reservoir exhibited perfect plug flow then the 'batch' models previously described should be appropriate for modeling the water quality within it. However, few storage tanks exhibit plug flow and in some tanks there can be pockets in which the turnover of water is very slow. In such cases, it is necessary to account for the variability in retention time throughout the tank. Considerable guidance on this is given in the AwwaRF report *Water quality modeling of distribution system storage facilities* (Grayman et al. 2000).

The internal surface area to volume ratio for a tank is significantly lower than that for pipes. Hence wall effects are likely to be much less significant for most water quality parameters. Nevertheless, it can still be important for microbial parameters.

Network Water Quality Models

The preceding 'batch' models predict the water quality of an individual vessel of water. However, to predict the water quality within a distribution network, it is necessary to account for the change in conditions imposed on a 'parcel' of water as it passes through the network. This can be achieved by combining the use of water quality 'batch' models with hydraulic network models, either by using:

- Separate network and water quality models
- Integrated water quality network models

Separate Network and Water Quality Models

For this approach network models are used to predict the water age at the point of interest. The water age is then used as an input to a batch type model. This assumes that the only input parameter for the batch model which varies throughout the network is age of water. For this to be valid either the water quality must be unaffected by interactions with the pipe surface, or all the pipes in the network should have similar characteristics.

In practice few networks exhibit the consistency required for a simplistic analysis and generally a more complex approach is required. The analysis needs to separate the different components of the network by similar characteristics and working from upstream to downstream analyze the change in water quality parameters within a component between discrete time steps. The resultant quality at the downstream end of the component will be the starting quality for the next section of pipes and time step.

This approach will inevitably be time consuming and the validity of the results will be subject to the extent of network subdivision, which itself will be restricted by the time available for the analysis. Since the starting point would be a validated hydraulic model, a more pragmatic approach for complex problems would be to use an integrated water quality network model.

Integrated Water Quality Network Models

These models calculate water quality as part of the network model algorithm. They will predict the analysis parameter at all nodes in the network and allow the results to be viewed graphically via the graphical interface of the network model.

The models work by tracking parcels of water as they pass through a network and calculate the changes in water quality as it passes through each link in the network. There are various methods by which this can be achieved or assessed based on similar algorithms to those used to model water age (previously discussed). In essence the models work as follows:

- The model performs calculations for a series of short time steps. The duration of these time steps may be fixed or vary depending on "defined rules".
- The network is divided into a vast number of small parcels of water. Typically each pipe will be subdivided into several parcels.
- Within each time step, the conditions imposed on each parcel of water (e.g. pipe characteristics) are considered to be constant allowing the water quality for each parcel of water to be calculated at the end of each time step.
- At the end of each time step, based on the flows predicted by the hydraulic aspects of the model, the new location for each parcel of water is calculated.
- A new set of water parcels is set up. The initial water quality of those parcels being inferred by the final water quality of each parcel from the preceding time step.

The water quality models incorporated into network models need to be relatively simple. It is quite common to build models with in excess of 10,000 pipes and it is not practical to define a large number of individual performance characteristics for each model element. Furthermore, model users are restricted to using the water quality models which have been incorporated into the network modeling package being used. Many will only allow a single first order kinetic model, although some provide more flexibility:

Bulk decay kinetic model: As previously discussed, reactions may occur either in the bulk flow or at the pipe wall. Some packages will allow users to define separate reaction coefficients for these different reactions (Equations 3.9 or 3.10). Often the bulk decay will be a simple first order kinetic model. However, there are alternative models worth considering. The models available within the EPANET network modeling software are summarized in Table 3.2 (Equations 3.7, 3.12, and 3.18 to 3.20).

Wall decay kinetic model: Those network models which allow bulk and wall decay to be modeled separately will generally follow the EPANET wall decay model (Equation 3.10). This model incorporates the rate of mass transfer of the water quality parameter to the pipe wall. The rate of mass transfer is based on the molecular diffusivity of the parameter of concern. The user should be aware that the default value for molecular diffusivity within EPANET is based on free chlorine at approximately 18 °C (Edwards, Denny, and Mills 1976) which will be significantly different for alternative parameters or temperatures. EPANET also allows the wall decay to be modeled as zero order (i.e. a constant reaction rate with time) (Equation 3.21). Vasconcelos et al. (1997) found this to give improved performance at modeling chlorine decay in unlined iron pipes.

$$\frac{dC}{dt} = k_w \qquad (3.21)$$

Table 3.2
Bulk decay models available in EPANET

Equation		Comments
$\frac{dC}{dt} = k_b(C_L - C)C^{(n-1)}$	(3.18)	*EPANET bulk decay equation:* It can be set as several well-known kinetic models by adjusting the value for C_L, k_b, and n (reaction order)
$\frac{dC}{dt} = k_b C$	(3.7)	*First order decay:* ($k_b < 0$, $C_L = 0$, $n = 1$): Widely used for chlorine decay.
$\frac{dC}{dt} = k_b(C_L - C)$	(3.12)	*First order saturation growth:* ($k_b > 0$, $C_L > 0$, $n = 1$) Typically used to model THM growth
$\frac{dC}{dt} = k_b C(C - C_L)$	(3.19)	*Two-component, second order kinetics:* ($k_b < 0$, $C_L \neq 0$, $n = 2$) Can be used to model the decay or growth of a compound when the reaction rate decreases as the concentration of the compound to which it is reacting decreases. May be applicable to chlorine reactions between chlorine and NOM.
$\frac{dC}{dt} = \frac{k_b C}{C_L - C}$	(3.20)	*Michaelis - Menton kinetics:* ($k_b \neq 0$, $C_L > 0$, $n < 0$) This rate equation is often used to describe enzyme-catalyzed reactions and microbial growth. It produces first order behavior at low concentrations and zero order behavior at higher concentrations.

Source: Adapted from USEPA 2002.

Decay coefficient modifier: Some modeling packages allow the decay coefficient to be defined as a function of other parameters. Such relationships include:

- There is evidence to suggest that the same processes that increase a pipe's roughness with age also tend to increase the reactivity of its wall with some chemical species, particularly chlorine and other disinfectants. EPANET allows each pipe's wall reaction coefficient (k_w) be a function of the coefficient used to describe its roughness.

$$k_w = -\frac{w}{\log\left(\frac{\lambda}{D}\right)} \tag{3.22}$$

where w = wall reaction pipe roughness coefficient
λ = Darcy-Weisbach friction factor

- Virtually all chemical reaction rates increase with temperature. A widely used rule of thumb is that the reaction rate doubles for every 10 °C rise in temperature. A more precise prediction of the relationship between reaction rates and temperature is given by the widely accepted Arrhenius equation (Coulson and Richardson 1964).

$$k = F \exp\left[-\frac{E}{R}(T + 273)\right] \quad (3.23)$$

where F = frequency factor (units dependant on order of reaction)
E = activation energy (J/mol)
R = Ideal gas constant = 8.31 J/mol°C
T = temperature (°C)

Chlorine

By far the most common water quality parameter to be modeled using network models is chlorine. As previously discussed, there has been much debate about the benefits of various kinetic models for chlorine decay models, but most network modeling software assumes first order decay. This has the merit of simplicity and due to its wide-scale use, most existing data on chlorine decay rates has been presented in terms of first order decay coefficients.

A key choice for the user is whether to model bulk and wall decay separately or use a single decay coefficient to describe both. Although separating bulk and wall coefficients may appear to increase the complication, there are many benefits in doing so:

- The bulk decay coefficient can be directly measured for each source water via a jar test (see previous). Hence it can generally be defined with reasonable confidence.
- By considering wall decay in isolation, it is more likely that the decay coefficient can be characterized allowing the user to assume decay coefficient to be similar to decay coefficients found from previous studies
- When applying the models it allows the user to consider what would happen if changes were made to the source water quality or pipes are renovated. For example:
    - The commissioning of granular activated carbon (GAC) treatment might be simulated by reducing the bulk decay coefficient
    - Pipe renovation can be modeled by reducing the wall decay coefficient
- It allows the model to be updated in response to changes in parameters such as temperature, organic content, or initial chlorine concentration

While numerous studies have shown models to be capable of being calibrated to reproduce the observed network conditions for a given day, a major limitation to modeling chlorine is that the decay coefficients, in particular the bulk decay coefficients, can show significant seasonal variation. This is because the rate of decay is influenced by a number of parameters which may vary significantly with time, but which are not included in the first order kinetic model. In particular, Powell et al. (2000b) has shown the bulk chlorine decay to be controlled by temperature, natural organic matter (NOM), and the initial chlorine concentration. These can be accounted for in a number of ways:

- Identifying relationships with the decay coefficients and updating them accordingly
- Measuring the bulk decay coefficient on a regular basis and using an appropriate value in the model
- Monitoring chlorine levels at a strategic point in the network and factoring all the decay coefficients up or down until the model gives a reasonable match

THM

Modeling THMs was one of the first applications of water quality models (Males, Grayman, and Clark 1988, Grayman, Clark, and Males 1988, Clark, Grayman, and Males 1988). In these models it was assumed that THMs reached their maximum concentration at the treatment works. They modeled THMs as conservative substances analyzing the blending of water from multiple sources with different THM concentrations. The models had limited point prediction capability, which was partly attributed to the assumption that the THMs were conservative and partly attributed to extreme fluctuations of THMs in the source waters.

Improved model performance has been achieved by simulating THMs as first order growth to a limiting value (Equation 3.12). Values for the coefficients to be used in each model were obtained via bottle tests (Vasconcelos et al. 1996).

The growth of THMs is generally modeled as a bulk water phenomenon that neglects the interactions at the pipe wall which, as previously discussed, may either increase or decrease THM concentrations. This approach makes THMs much simpler to model than chlorine as the reaction coefficients apply internationally to the model and can be measured directly via a jar test with no need to calibrate separate coefficients for each pipe. This also means that accuracy will not be impaired where the THM model is not integrated into the network model. Seidal et al. (2002) described such an approach based on predicting retention times at key points and then predicting retention time using the USEPA WTP model. The purpose of the work was to identify representative compliance monitoring sites as part of the IDSE as required by the Stage 2 DBPR.

Conservative Compounds and Blending

The concentration of some water quality parameters will not grow or decay within distribution. Nevertheless, it can still be useful to model them using network models as it allows the user to investigate:

- The propagation of fronts of high or low concentrations
- The blending or dilution of concentrations within the system

These applications do not rely on a water quality model, although they use the same age of water calculation algorithms. Modeling conservative substances has the following applications:

- Identifying the source of contamination in multiple source networks (Clark et al. 1991, Geldreich 1991). The results of such studies have been used in legal proceedings (Clark 1991).
- THM modeling – see previously
- Blending of nitrates
- Investigating discoloration problems caused by the mixing of water from two different sources
- Operational management of pollution incidents, but only if the models include all pipes and have been maintained in a calibrated state

Microbial

The developers of the PICOLLO network modeling package have developed an add-on module for their software, called PICCOBIO, which models the suspended biomass through a distribution system (Piriou et al. 1997, Gagnon et al. 1997). Few details of the underlying equations are available, but the following mechanisms are apparently incorporated:

1. The influence of BDOC on suspended and fixed biomass
2. The influence of temperature on bacterial dynamics
3. Bacterial mortality due to age and grazing
4. Bacterial mortality due to chlorine, taking into account the effect of pH on disinfection
5. Deposition and detachment of bacteria to and from the fixed biomass

The developers of the model claim a good correlation between modeled and observed viable bacteria counts ($R^2 = 0.80$) when applied to a French distribution network. The authors are not aware of any other published results from the use of this model, but understand that application in the UK has been less successful.

Further models for predicting microbial water quality or the risk of microbial failure are currently under development by other parties, but there are few published details to date.

TOOLS FOR PREDICTING WATER QUALITY CHANGES WHICH DO NOT REQUIRE CALCULATING RETENTION TIME

Strategic On-Line Sensors

There is an increasing availability and use of sensors which can record various water quality parameters within distribution. Although it is clearly better to measure a parameter directly than to try and predict its value, it is only feasible to monitor at a limited number of locations and there is likely to be considerable variability in water quality elsewhere in the network. Modeling techniques could potentially be used to predict that spatial variation.

Network water quality models have often been found to give reasonable predictions for a given day, but then rapidly fall out of calibration. This is due to temporal variation in the factors which affect the model coefficients. If models are used as a method of spatially interpolating data between given sample points, this problem is avoided. This could be achieved by developing methods for recalibrating model coefficients based on the current available data. Network modeling software is available which can import data from telemetry systems and optimize the model coefficients so as to best fit the observed network behavior.

Combining this updating procedure with regulatory sampling could significantly enhance the interpretation of regulatory data. Regulatory monitoring is inevitably limited and there may only be one or two chlorine samples taken from each network per week. However, in many networks it is possible to find regions of high or low chlorine concentration, making it difficult to interpret results. Network models could effectively be used to interpolate that data to give a full temporal and spatial picture. Models could also be used to quantify the performance, for example by calculating the percentage of the demand containing less than 0.05 mg/l and thereby allow direct comparison between the performance of different networks. By combining this with demand and temperature forecasting, it is potentially possible to give warning alarms before a failure occurs. Given the increasing trend to integrate network modeling with GIS, this is becoming feasible for many water utilities.

Tools used to predict water quality are based on an understanding of the changes that occur in water quality within a distribution system. While retention time may be a major factor in contributing to the change it is not necessarily required to make predictions. In some cases retention time may be important but other factors dominate in their effect on water quality. Knowledge of water quality changes can be derived from the system itself (either through experience, sampling, or historical data analysis) or can be based on a generic knowledge that applies to a range of networks.

Review of System Behavior

The tools in preceding sections are based on rules or relationships that may involve some degree of calibration with respect to the system to which they are derived. An alternative approach is to base the analysis purely on data collected from the system of concern. This can help the operator understand the behavior of that system and may allow predictions to be made about its future behavior.

Statistical Models

A number of attempts have been made using statistical models to predict the likelihood of microbial failures. Where sufficient data is available, it is possible to demonstrate statistical associations between the factors that affect water quality in distribution systems.

LeChevallier et al. (1991) demonstrated through statistical analysis that coliforms were associated with a number of factors including nutrients (AOC), temperature, disinfectant type, and disinfectant residual concentration. They also found that a model produced by multiple linear regression based on free chlorine residuals present in the dead end sections of the distribution network and temperature predicted 83.8 % of the variation in HPC bacterial numbers. The model predicted that at 20 °C, a free chlorine residual of 1 mg/L would be necessary to maintain HPC below 500 cfu/mL.

Volk and Joret (1994) developed a simple and logical model to identify circumstances where there is an appreciable risk that coliforms may occur in drinking water. The model incorporates four water quality parameters: temperature, bacterial density (as measured by epiflorescence), free chlorine residual, and BDOC. For each of these variables, graphical analysis of the number of coliforms in samples against the respective measurement of the water quality parameter was used to derive a 'threshold' defining the range in which coliforms can be present. The four thresholds then provide a basis for scoring each sample on a five point scale, according to whether 0, 1, 2, 3 or 4 of the threshold values have been violated. This score gives an assessment of risk, which the authors link informally to the observed occurrences of coliform-positive samples. Sampling sites are described as 'exposed', 'weakly exposed', and 'without risk' according to whether the number of thresholds exceeded by the water sample is 4, 3 or 2, 1 or 0, respectively.

Gatel et al. (1995) called this a "black box" model for predicting coliform occurrences based on threshold values of the explanatory variables: temperature, bacterial density, free chlorine residual, and BDOC. If any of the explanatory variables exceeds its threshold value then the model increased a qualitative risk variable to a higher state.

Gale et al. (1997) used a statistical approach based on the concept of "generalized linear modeling" (GLM) to develop coliform predictive models. Their aims were to:

1. Identify statistical associations between parameters measured as part of the UK statutory water quality monitoring program
2. Develop and validate models for predicting coliform occurrences
3. Demonstrate the application of the models in maintaining and improving water quality

The results of the study show that coliforms were associated with higher numbers of heterotrophic bacteria, temperatures, and lower total chlorine residuals. Associations were also shown with coliforms and color and turbidity. Models were calibrated by GLM to predict the probability of detecting coliforms in statutory samples from specific groups or zones within UK water companies. Validation exercises indicated that the models were suitable for cautious application within the zones in which they were calibrated and in related zones within the same company. Figures 3.9 and 3.10 show the output from the models.

The models can be used for predicting coliform occurrences after changing the associated water quality parameters. Figure 3.9 shows the relation between coliform occurrences and temperature and indicates that the operator should consider remedial taking action when temperatures exceed 15 °C, such as deliberately drawing down storage levels to turnover stored water. In Figure 3.10 the relationship between coliforms and chlorine is shown. In this case the operator should maintain chlorine residuals above 0.2 mg/L in all parts of the system.

A limitation of this approach is that all data is aggregated to zonal statistics before looking for correlations. Individual extremes in any of the data may be directly attributable to extremes in related water quality parameters for a given sample. For example, a coliform failure may be directly related to the high temperature of a sample. However, by aggregating the data together, the significance of these extremes in the final model is diminished.

The success of statistical modeling is dependent upon the quality and quantity of the data analyzed. The small number of coliform positive samples (<1 %) used in Gale's (1997) study make the models highly susceptible to both statistical and sampling errors. However, statistically significant results were produced with a positive influence on coliform failures produced by increases in temperature and HPC and decreases in total chlorine.

Data Analysis

A utility's business management information system (MIS) contains a wealth of physical system and operational performance information, all of which can be assimilated for analysis if necessary. The MIS may comprise a single data management system or a collection of electronic databases and paper records.

Any analysis of distribution network water quality will generally require information from some of the following sources:

- Record drawings of the distribution network (GIS database or paper copy) containing:
    - Configuration of the network and ancillary equipment
    - Hydraulic boundaries and water quality zones
    - System hydraulics and operation and control regime

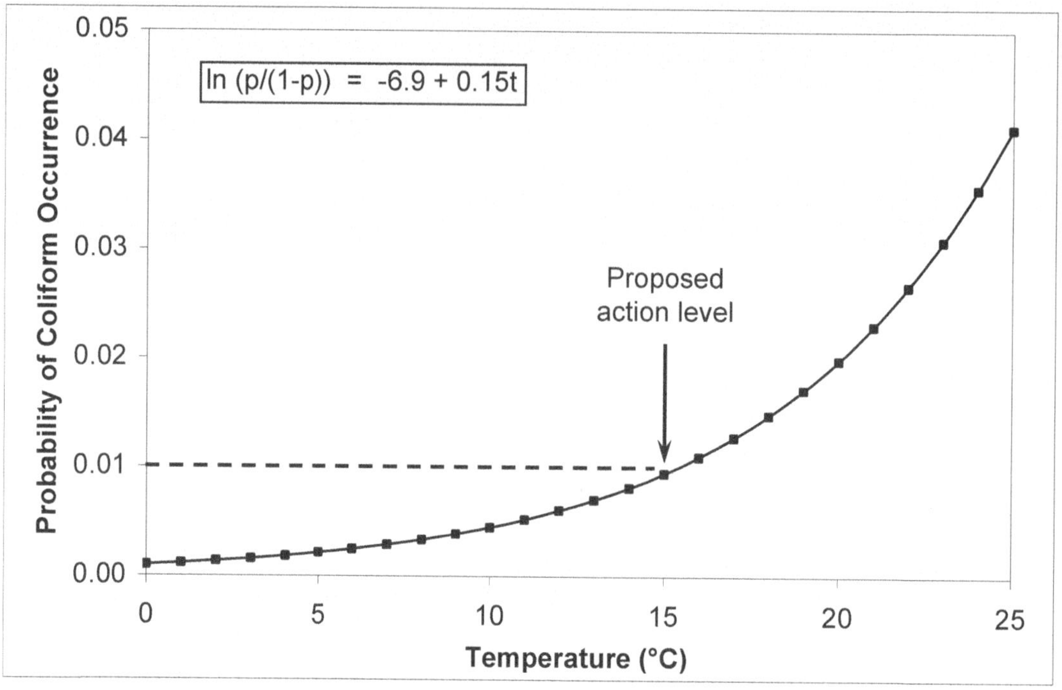

Figure 3.9 Coliform predictive model 1
Source: Adapted from UKWIR report 97/DW/02/3 (WRc 1997b)

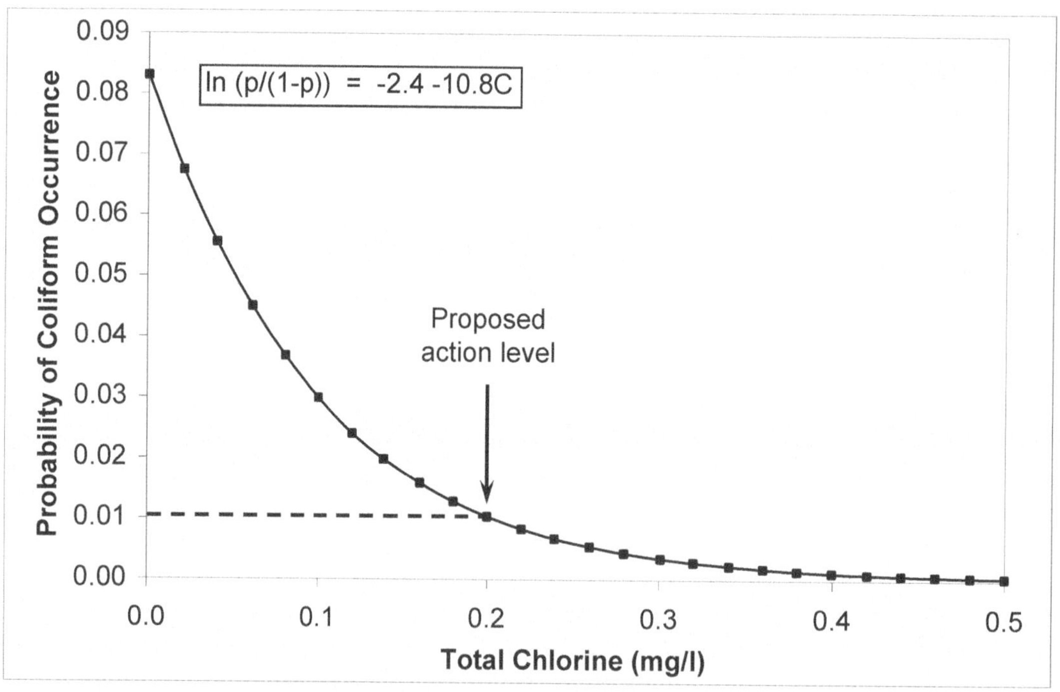

Figure 3.10 Coliform predictive model 2
Source: Adapted from UKWIR report 97/DW/02/3 (WRc 1997b)

  - Mains materials, condition, and maintenance history (bursts and repairs, ground conditions, etc)
  - Areas and details of network rehabilitation in previous years
  - Records of planned network rehabilitation, flushing, and maintenance
  - Specific plans for improving water quality
  - Known system problems particularly with reference to water quality
  - Storage tank details; internal layout, and inlet and outlet arrangements
- Hydraulic network and water quality models where available, including a record of network changes since the model was calibrated
- Fixed and random location sample data from:
  - Source waters and treatment works
  - Within the distribution network and at storage facilities
  - Locations of chemical bacteriological sample points
- Sundry data:
  - Telemetry data or system monitoring records
  - Leakage and pressure management data
  - System and ancillary equipment maintenance records
  - System performance data for regulatory reporting
  - Customer database and meter reading records
  - Customer complaint records and Customer tracking surveys
  - District inspectors reports

Much of this information can now be held electronically and linked to GIS for visualization in order to aid assimilating the information and testing relationships between different data sets. Using GIS to analyze and display data is discussed later in this section.

The quality of a data set will determine the level of confidence in any analysis which uses that data. Before starting an analysis, it is important to understand the significance of the output and hence actions that may be taken based on the accuracy of the raw data. Although this also applies to the accuracy of the analysis tools, generally limitations of the input data rather than the quality of the analysis tool will determine the overall quality of any output from a model.

Customer surveys: There are three ways customers can be canvassed for information about the quality of the service that they receive. The most common interface with the customer is when he is registering a complaint. Although valuable information can be derived from the exchange, the responses from the customer tend to be biased and can represent a relatively low quality of information. Its quality can be improved both by implementing a utility wide standard complaints recording procedure and correlating the complaints against other company performance and water quality analysis data. The cause of the initial complaint may be a system problem manifesting itself at the customers tap, or problem related to the supply to that customer.

Response surveys result from a visit to a customer following a complaint or request for a visit to the premises by a district inspector to investigate a supply problem or to assist the customer with a repair or maintenance action. During such visits, it is common practice to question the customer about the quality of service. Where, during the visit, flow and pressure are monitored for levels of service and a water sample is taken for analysis, the survey responses can be correlated against the level of service measurements and analysis results both at the customer's premises and within distribution. Random survey data tends to be more selective than customer complaints data, primarily because the report will only be closed when the problem is resolved and because it only represents problems within a customers premises.

Customer tracking surveys are usually undertaken on an annual basis to monitor changes in the public's perception of the service they receive. They are often undertaken by an independent survey company. It is important to ensure that sufficient customers are surveyed to give statistically significant results which can be difficult for small areas. The questions should be carefully devised to eliminate preconceptions. They usually include taking samples and can include taste perception tests. Correlation analysis can be carried out on similar or related questions to test the consistency of answers.

GIS Data Visualization

GIS are used to store infrastructure data such as pipe characteristics (length, diameter, material, pipe age, and pipe frictional coefficient) and regional feature data such as soil characteristics (corrosivity and fracture potential) and map information (background map image, photographs, road networks, and building heights).

The water age and other water quality parameters can be attached to the pipe database. The data is therefore displayed as a sequence of images, each of which is extracted from the working database being updated at each time step.

Most of the current hydraulic models allow the user to carry out network tracing and water age mapping. For many simple distribution networks the actual location and the map of the area can be overlaid on the networks. Since operators need to have access to the data the output needs to be representative of the system. However, the visualization of current hydraulic networks software trails behind the fast moving GIS technology and so it can be beneficial to link the simulation results to the GIS database for visualization purposes.

Investigation of real networks requires their exact location. A gazetteer tool should be provided to locate an address. National mapping data can be used to automatically identify the geographic co-ordinates of customer addresses. GIS software provides tools to display additional information of the region and the user digitized information. This feature is essential for event response.

GIS data of the network is in non-simplified form and gives the exact location in the real world. A GIS data set is therefore much larger than the hydraulic model data. Handling the large data requires database manipulation tools consisting of data import and export tools, data linking and joining tools, and spatial linking and joining tools.

Figures 3.11 and 3.12 give examples of GIS data visualization. There are three exercises involved in preparing the network data for visualization:

1. *Network tracing*: This is used to mark out the water from various sources. The tracing can be directional and is used to determine the span of contamination or supply from a single source. Back tracing is also used to locate the source from two affected locations. Directional tracing requires the direction of flow to be known, however, if this is unknown, it may be inferred from pipe diameters by assuming that the flow is from large to small diameters. A network tracing exercise does not require the actual flow values. Tracing to locate the upstream source is carried out by performing the intersection of two randomly selected routes from two downstream locations, and then by calculating the success probability.

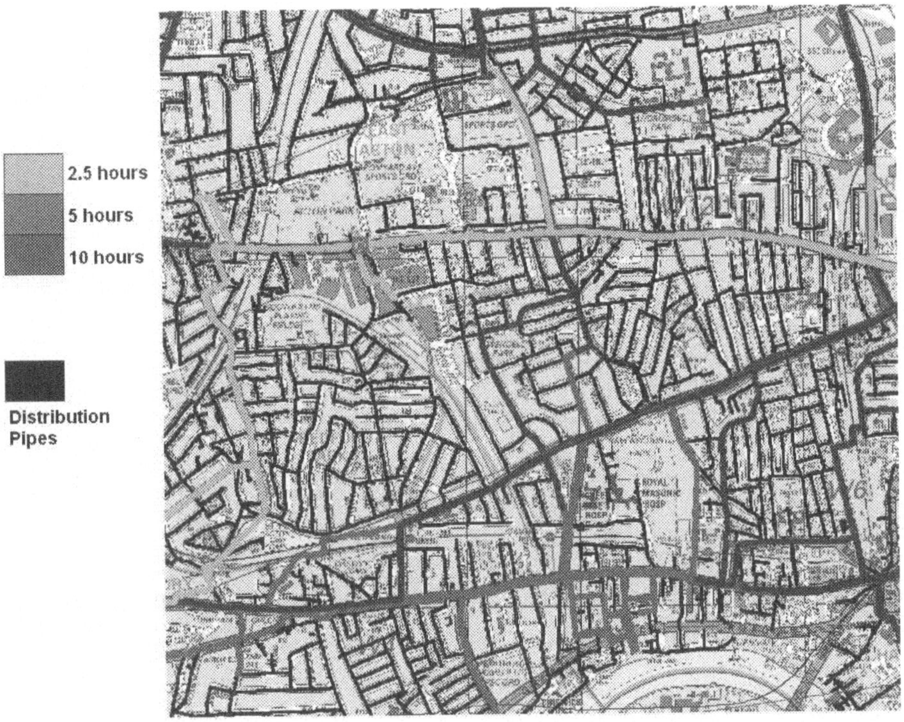

Figure 3.11 Network overlaid on street map of the area

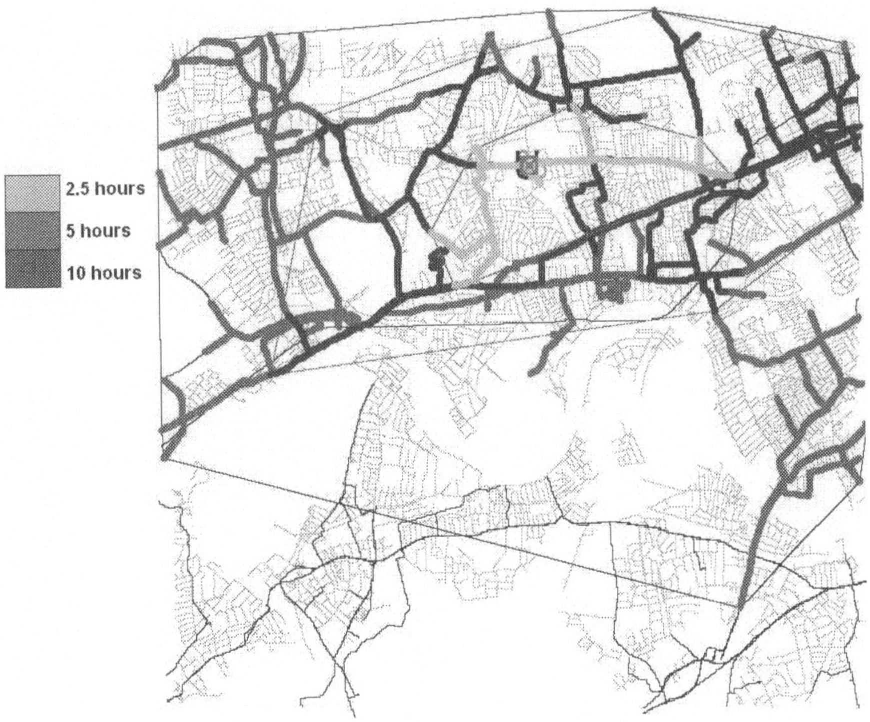

Figure 3.12 Network tracing from a service reservoir – contour of water age

2. *Network routing*: is used to determine the flow path satisfying certain constraints. This is normally an optimization process, such as finding the shortest distance or shortest time route between two locations. Network routing explicitly requires the network flow data. In network management, the network routing is used to determine the best possible routing to a hydrant in a flushing exercise.
3. *Network allocation*: is used to map out the effect from a water source. Figures 3.11 and 3.12 give examples of a water age contour. Network allocation explicitly requires flow data, although this data is often estimated. The allocation is obtained by tracing from the source subject to the distance or the time limit.

For large networks, the tracing process can be used to establish supply areas associated with service tanks. The smaller diameter pipes (< 100 mm) within an area are linked to nodes on the larger strategic pipes (200 mm diameter and above) which form the network.

Generic Rule Based Models

One of the goals of researchers is to produce a generic model that can be applied to all distribution systems. In many cases these were based on detailed research on one particular system. However, during the 1990's a number of researchers created tools or "rules of thumb" from reviewing literature and expert opinion in an attempt to derive generic models that can be applied to many, if not all, distribution networks.

Rules of Thumb

Various researchers have developed guidelines or rules of thumb to predict water quality changes in distribution networks. In the late 1980's and early 1990's there was a considerable amount of research into the factors that would lead to microbial growth in distribution systems. Studies were often carried out in the laboratory and the findings validated by monitoring in distribution systems. The key factors were shown to be nutrients (as measured by AOC or BDOC) temperature, and disinfectant residual (van der Kooij 1992, Block et al. 1993, Volk and Joret 1994, Huck 1990, LeChevallier et al. 1991). These researchers identified the limits for nutrients, temperatures, and disinfectant residuals required to support microbial growth (Table 3.3) and subsequently proposed models for the water quality changes that occurred in distribution systems.

Water temperature is perhaps the most important rate controlling parameter in the microbial growth process for drinking water. Most investigators have observed significant microbial activity in water at 15 °C or higher. Fransolet, Villiers, and Masschelein (1985) found that water temperature influenced not only the growth rate, but the lag phase and cell yield of the bacteria. For *Pseudomonas putida*, the lag phase was of the order of 3 days at 7.5 °C but only 10 hours at 17.5 °C. This suggests that at low temperatures bacteria could pass through the distribution system before significant growth could be achieved. They also found that the growth of the coliforms *E. coli* and *Enterobacter aerogenes* was very slow at temperatures less than 20 °C.

More recent models also incorporate the effects of the pipe wall materials on water quality as demonstrated by Colbourne and Brown (1979), Colbourne (1985), van der Kooij and Veenendall (1993), and Holt et al. (1998). Examples of some of the models for predicting microbial growth based on nutrients, temperature, and disinfectant residuals are discussed below.

Table 3.3
Limits for the factors that affect microbiological growth

Parameter	Reference value for biostability	Reference
Nutrients (AOC)	AOC ≤ 10 μg C/L	Van der Kooij, Visser, and Hijnen (1982), Van der Kooij (1992)
Nutrients regrowth potential	Growth rate ≤ 0.1 h^{-1}	Werner (1985), Hambsch, Werner, and Frimmel (1992)
Nutrients (BDOC)	BDOC ≤ 0.1 to 0.2 mg/L. Based on BDOC decrease in distribution systems.	Joret and Levy (1986), Servais, Billen, and Hascoët (1987)
Nutrients (AOC-ATP)	No value defined	Stanfield and Jago (1987)
Nutrients (rapid BDOC)	BDOC ≤ 0.1-0.2 mg/L	Lucena, Frias, and Ribas (1990)
Nutrients (rapid AOC)	AOC ≤ 50 μg C/L to prevent regrowth of coliforms	LeChevallier et al. (1991 and 1993)
Nutrients (biofilm formation rate (BFR)	Limited regrowth of *Aeromonads* when BFR < 10 pg ATP/cm²d	Van der Kooij and Veenendaal (1993), Van der Kooij et al. (1995 and 1999)
Temperature	Demonstrated that low temperatures caused significant delay to bacterial growth	Fransolet, Villiers, and Masschelein 1985
	> 16 °C increased bacterial growth	Dukan et al. (1996)
	> 15 °C increased bacterial growth	Chambers et al. (2000)
	Probability of coliform detection increased with increasing temperature (basis of predictive model)	Gale et al. (1997)
	Coliforms grow slowly below 20 °C	Fransolet, Villiers, and Masschelein (1985)
	Occurrence of coliforms more likely at temperatures >15 °C	LeChevallier, Schultz, and Lee (1991)
Disinfectant (chlorine)	Occurrence of coliforms in when temperatures >15 °C and monochloramine residuals <0.1 mg/L	Colbourne et al. (1991)

Unified Approach

Van der Kooij et al. (2002) developed the "Unified Biofilm Approach" for assessing the biological stability of water to enable the appropriate design of water treatment works and distribution systems, to control regrowth especially when disinfectants are not used. In the Netherlands where water is distributed with either low or zero disinfectant residual, the Unified Approach allows the operator to predict any regrowth potential. The approach is based on three observations:

- Assessment of the biofilm formation rate of water leaving the treatment facility in addition to determining the AOC concentration. This is determined by allowing biofilms to develop on glass rings in a biofilm monitor and then measuring growth.

- Assessment of the biofilm formation potential of synthetic materials in contact with drinking water. This is determined by measuring the concentration of active biomass on pieces of pipe material incubated in slow sand filtrate at 25 °C.
- Assessment of the biofilm concentration on pipe walls. This is determined from observations of biofilm growth on pipes at the treatment works or in distribution.

The amount of biofilm growth derived from the three tests is measured using adenosine triphosphate (ATP) to determine biomass and compared directly. These observations allow a ranking of waters and materials and together they proved a framework for predicting biostability in the distribution system.

"Toolbox" Approach

In 2000 UKWIR reported on their development of "toolboxes" for seven key drinking water parameters namely: (1) disinfectant decay, (2) trihalomethane formation, (3) iron discoloration and particulates, (4) tastes and odors, (5) biofilms, (6) coliforms, and (7) pH, based on literature searches and extensive consultation with experts in the industry (Chambers et al. 2000). Each toolbox provides instructions on how to use, background information, identifies facts about the parameter and contains rules of thumb to assist operators managing distribution systems and treatment works to maintain or make improvements to drinking water quality.

The "toolbox" for trihalomethane formation is presented in outline to provide an example of the approach taken and the information contained in the report. The aim of the THM toolbox is to enable planning of interim actions where treatment cannot be improved in the short term or where improvements have not solved existing problems. It also enables the operator to identify and characterize problem water sources. The main facts are identified as:

- The most important method for controlling the formation of THM is to ensure that adequate treatment is applied prior to distribution
- The key factors to consider for controlling the formation of THMs are:
  - The concentration and nature of organic material in the water
  - Free chlorine dose, free chlorine residual
  - pH
  - Temperature
  - Contact time
  - Bromide concentration
  - Inorganic chlorine demand
  - Monochloramine
  - Chlorine dioxide

Based on the above facts and from information gathered in the literature and the expert review a number of rules of thumb were provided:

- At low water temperatures (less than 10 °C) THMs concentrations in distribution do not increase significantly. Therefore, chlorine should be controlled or TOC removal should be enhanced whenever the water temperature rises above 10 °C (e.g. by reactivating the GAC filters at the beginning of summer).

- For treated water TOC values of less than 2 mg/L, THMs are unlikely to exceed 80 μg/L
- For TOC values greater than 4 mg/L it will be difficult to prevent THMs from exceeding 100 μg/L if free chlorine is maintained to the tap for travel times of 2 to 3 days
- Where the source water is highly colored, THMs can exceed 100 μg/L even though the water is treated such that the color is reduced to an acceptable level
- If optimizing treatment does not enable THM concentrations of less than 80 μg/L to be achieved then the use of chloramination should be considered
- Increasing the chlorine dose at entry (or at booster points in the distribution system) will increase the final concentrations of THMs
- Different models have appreciably differing power relationships with respect to chlorine dose but the exponent is always less than 1, typically 0.5 or less. A very rough rule of thumb would assume the dose exponent to be 0.5, the increased THMs could then be estimated from:

$$[THM]_1 = \frac{([Cl_2]_1)^{0.5}}{([Cl_2]_0)^{0.5}} \times [THM]_0 \qquad (3.24)$$

where $[THM]_1$ = THM concentration given by new chlorine dose
$[Cl_2]_1$ = new chlorine dose
$[Cl_2]_0$ = initial chlorine dose
$[THM]_0$ = THM concentration given by initial chlorine dose

For example, if the old dose was 2.0 mg/L and this gave rise to 80 μg/L THMs at the tap then a new dose of 2.5 mg/L would result in approximately 90 μg/L of THMs.

The authors also identified a number of complex formulations, with up to 8 parameters, that have been developed to predict THM formation (primarily in treatment). These are probably over complicated for predicting changes in distribution systems. As an example, a schematic diagram is given (Figure 3.13) showing the relationship between TOC, pH, temperature, and the formation of THMs for a given chlorine dose and a given contact time. Each line represents the formation of 100 μg/L THMs at a given temperature. The area above a line indicates that THMs will be greater than 100 μg/L, the area below a line indicates that THMs will be less than 100 μg/L. The diagram is based on a model of Amy, Chadik, and Chowdhury (1987), as used in Sohn et al. (1998).

The importance of taking an holistic approach to improving water quality is emphasized in the report. While each of the seven parameters have individual *"toolboxes"*, the application of rules from one toolbox can have an impact on more than one parameter. In some cases the impact might be small, in others it might be substantially positive, and in some it might be substantially negative. Where the latter is apparent then it may be necessary to use an alternative approach. It might also be possible to accept the negative impact on the basis that the benefits from other aspects outweigh the disadvantages.

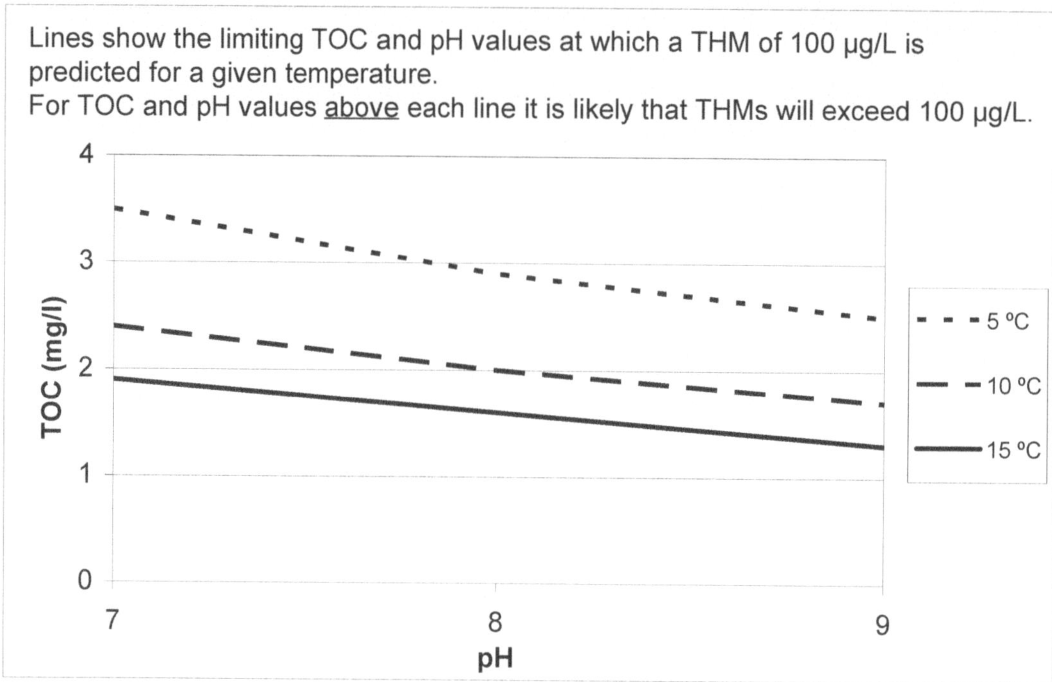

Figure 3.13 Relationship between TOC, pH, temperature, and THM formation

Table 3.4 is reproduced to show the extent of the interactions between the various water quality factors. In the table biofilms and coliforms are represented under the heading of microbiological quality.

Empirical Models

Chemistry Dissolution Models

There are several software packages that allow the prediction of metal dissolution dependent on water quality characteristics. For the main metals of concern, from a health and regulatory perspective, i.e. lead, nickel, copper, and chromium, the dominant factor affecting dissolution of the metal is the alkalinity of the water. While retention time is important in governing dissolution in consumers' premises, it is not considered a "network" issue. However, AwwaRF and others have produced software that allows the prediction of dissolution of metals based on water quality and retention time. Examples for corrosion control are the Rothberg, Tamburini and Winsor (2000) Model for corrosion control and process chemistry (Version 4.0), the Kiwa model for softening and pH control, AQUACALC, and the model developed by Kippin et al (2001) for blending of waters, MULTIBLEND.

Table 3.4
Impact of applying the rules of thumb

	The impact of maintaining drinking water quality through						
	Reducing disinfectant decay	Reducing THMs	Reducing biofilms		Reducing iron and discoloration	Reducing chlorine taste and odor	Increasing pH to reduce metals etc.
Main actions to reduce problem*	Reduce travel time, clean pipes, replace corroded iron	Reduce travel time or use chloramine	Reduce AOC	Increase chlorine§	Modify operations, flushing progs., remove or rehab. corroded iron pipe	Reduce chlorine or variability of concentration	Increase pH and/or hardness or alkalinity
Impact on disinfectant decay	Improvement	Improvement	Improvement	Deterioration§	Improvement	Improvement	Improvement†
Impact on THMs	Improvement	Improvement	No effect	Deterioration	Small improvement	Improvement	Deterioration
Impact on microbiological quality	Small improvement	Improvement	Improvement	Improvement	Improvement	Deterioration	No effect
Impact on iron and discoloration	Improvement	No effect	Unpredictable‡	Small deterioration	Improvement	Small improvement	Improvement
Impact on taste and odor	Small improvement	Improvement	Small improvement	Deterioration	Small improvement	Improvement	No effect
Impact on pH	No effect	No effect	No effect	No effect	Small improvement	No effect	Improvement

Source: Adapted from UKWIR report 00/DW/03/11 (Chambers et al. 2000)
* Typical actions are listed
† Although the disinfectant decay is reduced, the disinfection power is also reduced which is detrimental
‡ Effect is unpredictable: could reduce protective film and hence worsen discoloration or stop anaerobic corrosion under the biofilm which would reduce discoloration
§ Rate of deterioration will increase

Artificial Neural Networks

An artificial neural network (ANN) is an empirical model which learns from representative historical data, rather than being programmed with specific laws. In so doing, it can discover complex non-linear relationships between input and output variables, accommodating subtle intricacies that are overlooked when the problem is reduced to relatively simple mathematical functions.

Water companies increasingly collect vast quantities of data about the behavior of their networks for operational and regulatory purposes, much of this data is likely to be related, but that relationship may be hard to define. It has been argued that ANNs are likely to be well suited to applications within water treatment and distribution (Rodriguez and Sérodes 1994). While programming the models with data is in principle straightforward, a considerable level of expertise is required to control the fundamental structure of the model. Unless software packages are developed which include predefined model structures for application to water distribution problems, they are unlikely to become a routine tool. While ANNs are capable of identifying complex relationships, they may struggle where there is a high degree of inaccurate or redundant data. Hence ANNs are probably most suited to process control rather than generalized analysis of distribution systems. In particular they are likely to struggle when applied to microbial water quality when relationships with other parameters result in an increased risk rather than a pre-determinable value.

SUMMARY

A considerable amount of research has been carried out, much of it under laboratory or pilot scale conditions, which has increased our understanding of the relationship between retention time and water quality changes. This has allowed the development of two approaches for predicting water quality changes, one based on calculating retention time and the second based on empirical knowledge and water quality data. These two approaches have led to the development of a number of tools for the analysis of water quality in distribution networks. Summary details of the main tools available are presented in Table 3.5.

Table 3.5
Summary of retention time analysis tools

Method	Details
Manual calculations	Simple systems can be analyzed using manual and spreadsheet calculations. Pipe flow equations combined with storage or turnover assessments can be used to derive retention time estimates for snapshot conditions and simple system extended periods. Age is then compared with water quality records and customer complaints to identify parts of systems prone or liable to quality failure (chemical and biological).
Rules of thumb and experience	Generic rules are used together with hydraulic assessment and an understanding of system operations (e.g. for a specific Cl_2 dose, expect stagnant water complaints where age is over 100 hours). Operators' practical and specific experience can be used to identify parts of their system prone to water quality problems (e.g. under certain demand or supply conditions, problems may occur at known locations).
Tracer studies	By injecting a chemical into a network and then monitoring for it at a downstream point, the retention time can be directly measured. This technique is relatively straightforward, especially when the trace chemical is already present in the system and can be easily adjusted e.g. fluoride or chlorine.
Review of water quality data	Review of water quality data can indicate areas with water quality problems. These may then correlate with areas of high water age
Geographical information systems (GIS)	GIS can be used to select those properties which are at the extremities of the network or near dead ends and so vulnerable to a high water age. It is also possible to incorporate more sophisticated techniques by assigning characteristic water quality properties (including retention time) to each element in the GIS. Queries can then be set up which follow rules to estimate the water quality within the network. GIS can also be used to present database information spatially to look for patterns and causes of problems.
Hydraulic network model	Computer models of the hydraulic behavior of distribution networks such as EPANET, H2Onet, Stoner, WaterCad, Aquis, and Infoworks. This software can be used to predict the water age and highlight any areas in which it would exceed a given value.
Water quality network model	Many of the above network models can also be used to model water quality for parameters whose concentration is a direct function of water age. The most commonly modeled parameter is chlorine. However, this does require a reliable estimate of reaction coefficients that may not be readily available.
Water quality statistical model	Some utilities have developed models of water quality, which are purely derived from statistical analysis of the vast quantities of water quality data that are collected for regulatory purposes
Physical model	An expensive tool, not normally used for system operational management. However it may be useful for validating new design (e.g. the location of baffles to prevent dead zones in reservoirs).
Computational fluid dynamics (CFD)	CFD is increasingly being used to analyze the behavior of water within tanks and reservoirs. It can highlight dead zones in which there is limited turnover of water. Methods for rectifying problems such as installing baffles can also be modeled. CFD models have also been used to simulate flow in pipes including at dead ends under flushing conditions.
Reservoir model	Assessments of turnover including strategic storage, diurnal, and seasonal usage give indications of retention time and hence the preponderance of water quality problems. However three-dimensional modeling (CFD or physical models) may still be needed to understand internal circulation, dead zones, etc.
Customer perception surveys	Carefully designed and executed surveys can provide useful information and insight into particular problems. However, they need to be used in conjunction with other tools to provide information on retention time issues. They have been widely used in the UK in conjunction with structured sampling to identify and justify network improvement schemes.
Neural network models	Neural networks are potentially of benefit where it is reasonable to assume that there is a relationship between parameters, but where the mechanism by which they are related is not understood. They are effectively used to derive empirical relationships and could be used in a similar method to statistical water quality models (see above). To date they have not been widely used.
Optimization software	Optimization tools are widely available within spreadsheets (e.g. the 'Solver' add-in within Microsoft Excel) and can be used to optimize many problems that can be defined in terms of mathematical functions. Optimization tools are also increasingly being incorporated into other analysis software such as network models.

CHAPTER 4
METHODS FOR CONTROLLING RETENTION TIME

Many utilities are likely to have implemented some form of retention time management techniques without specifically classifying it as 'retention time management'. Much can be done to reduce retention time by relatively simple and inexpensive techniques such as flushing at dead ends or reducing the operational top water level of reservoirs. Other techniques will involve more significant capital works. This chapter outlines some of the methods that can be used to control retention time. It is intended to evaluate these methods through field trials in Phase II of this project and so the Phase II report will give greater detail about the methods, their benefits, and limitations.

Controlling retention time will affect other operational or service level requirements of the network. These should be considered prior to the implementation of the work to check that they will remain within acceptable limits. In particular, many of the methods will reduce the hydraulic capacity or storage within the network, an issue that needs to be balanced against fire flow and supply pressure requirements.

Altering the Valving of Networks

Retention times are often highest at dead ends or valves which have been shut to produce internal boundaries. It is possible to reduce retention time in localized parts of a distribution network by changing valve arrangements. Internal boundaries are necessary for many operational reasons including leakage monitoring and pressure control. However, retention time problems can be reduced by minimizing the number of shut valves required to produce the hydraulic boundary and locating the valves in streets with relatively high demand on either side of the shut valve.

In some circumstances shut valves can be beneficial. By inserting shut valves in a network flows can be rerouted through parts of the system with low demand and high retention time where otherwise the flows pass direct to points of high demand. Although the age of water may be increased at the point of high demand, other parts of the network will benefit from higher turnover.

Installing Time Varying Valves

It may be possible to use time actuated valves to reduce retention time. For example, if hydraulic boundaries are required for pressure control, they may only be required during periods of peak demand. Outside of these times, the boundaries could be left open. Time actuated valves can also be useful for controlling the flow through storage tanks.

Manual Flushing

Manual flushing is often used as the first remedial measure following a water quality failure (microbial, discolored water, or low chlorine). The primary purpose of flushing is to expel the contaminated water, but it also reduces the retention time and thereby improves the water quality temporarily. Some utilities have weekly flushing programs to control water quality in problem areas. Historically this was most frequently practiced to prevent stagnation in long dead ends which could otherwise result in red water incidents. In the USA following the introduction of the minimum disinfectant residual requirements of the SWTR, regular flushing is also used by some utilities to obtain a higher residual to meet the requirements of the SWTL.

Automated Flushing

A major limitation of using flushing as a means of controlling retention time is that the period between flushing must be less than the return period of a water quality violation event. For many water quality parameters this may require weekly or even daily flushing which is expensive in terms of manpower. To address this, automated flushing devices have been developed. These devices purge the system at regular intervals either by discharging water to a sewer or water course, or by discharging it over the surrounding ground. Some systems incorporate a dechlorination facility to prevent discharging chlorinated water to the environment.

The use of automated flushing has some drawbacks:

- The disposal of water may be difficult especially in urban areas or in freezing conditions
- The volume and value of wasted water may be unacceptable. However, due to improved control it may result in less water wasted than if manual flushing was used.
- The units may be prone to vandalism or tampering if installed in public areas

Automated flushing devices are best suited to rural networks in which security of the units and disposal of waste water is less problematic. In networks with long pipe runs terminating in dead ends there may be few viable alternatives to flushing for controlling retention time.

An alternative to automated flushing sometimes practiced is to have a continuous bleed of water to waste at a dead end. This will prevent stagnation of the water, but again there are problems with wastage and disposal.

Abandoning Mains

Many distribution networks have been steadily added to over many years. This invariably results in a non-ideal network. Often new mains will have been laid parallel to old mains to reinforce hydraulic capacity. If the new main is significantly larger than the old main then the old main may not be required in order to meet the increased hydraulic demand. However, if there are connections off the old main, it is likely to have been retained in service in order to avoid the expense of transferring the connections onto the new main.

Since the retention time in each stretch of pipes of a given material and condition is critical to water quality and not the absolute water age, abandoning a stretch of poor condition pipes may alleviate water quality problems even if it does not significantly reduce the overall water age in the network.

Downsizing Mains

Most networks have been designed to meet a minimum hydraulic capacity. Additional capacity is generally included at the design stage to accommodate for future growth or to allow more flexibility in the configuration of a network. A utility may also have a policy to limit the number of different pipe diameters within the system in order to simplify construction and maintenance. Consequently network pipes tend to be larger than is necessary to meet the daily demand from the network leading to increased retention times. Hence, there is often scope to replace mains with smaller diameter pipes but still achieve the required hydraulic capacity.

In the Netherlands, some utilities have started to design new networks specifically with the intention of minimizing retention time (See Chapter 1 Network Design). This has involved questioning the fundamental demand requirements used to design the network; in particular fire flows have been reviewed and fire hydrants are now generally located in through-mains in preference to dead ends allowing much smaller diameter pipes (down to 40 mm) to be used in the dead ends. A similar approach applied to existing networks would allow mains to be significantly downsized.

Increasing Turnover of Storage Facilities

Lack of water turnover in storage facilities has long been recognized as a primary cause of water quality problems within distribution. There are two aspects which need to be considered:

1. The retention time is directly related to the volume of storage. Hence, it is important to avoid unnecessary storage.
2. Ideally tanks should exhibit a fully mixed flow regime. In practice, this is rarely the case and there will be pockets of water which are not well mixed with the bulk of the water resulting in stagnant zones. The age of water in these zones can be considerably higher than the average age of water in the tank.

Guidance on the above is given in the AwwaRF report *Maintaining water quality in finished water storage facilities* (Kirmeyer et al 1999).

Adjusting Pump Schedules

One of the purposes of storage is to balance the variable demand for water from consumers with the output from single speed pumps which are either on or off. It is desirable for the pumps to run during periods of cheap energy. By adjusting the pumping regime, it is often possible to improve the balance between network demand and the supply from the pumps and thereby reduce the volume of storage required. Depending on electricity tariffs, a consequence of this method of control can be higher operational energy costs.

Reducing the Operational Top Water Level (TWL) of Storage Facilities

Where there are no supply pressure implications, reducing the operational top water level is a simple and effective method of reducing storage. This operational device can be employed on a seasonal basis, providing more storage in the summer period when the network demand is higher, or it can be reduced permanently.

Taking Storage Out of Service or Removing Tank Cells

If storage is not required then consideration should be given to removing it either by removing the entire tank or one or more cells for a multi-cell facility.

Altering Reservoir Configuration to Avoid Dead Zones

Storage tanks should be designed and configured so as to prevent areas of dead water. However, tanks with common inlet and outlets tend to only turn over water in the vicinity of the inlet and outlet connection. Fitting baffles into reservoirs can control the flow stream between the inlet and the outlet and locating the inlet and outlet at opposite ends of the tank can help to give adequate turnover in all parts of the tank.

SUMMARY

There are many techniques by which retention time within distribution networks can be controlled ranging from inexpensive operational measures to costly capital improvements. The principle methods are summarized in Table 4.1.

Table 4.1
Methods for controlling retention time.

Method	Details
Altering the valving of networks	By changing valve arrangements and hydraulic boundaries, travel times can be reduced and water rerouted to increase velocities in low flowing pipes to maximize velocities.
Installing time varying valves	Time and other control valves can be used to control flows.
Manual flushing	Periodic flushing to remove sediment and reduce water age in dead ends and low velocity sections of pipe.
Automated flushing	Automatic programmed flushing to remove sediment and reduce water age in dead ends and low velocity sections of pipe.
Abandoning mains	Remove surplus capacity from system.
Downsizing mains	Reduce system capacity to increase velocities and reduce retention times.
Increasing turnover of storage tanks.	Eliminate or reduce strategic storage and manage diurnal storage to minimize retention time. This is likely to be linked to resource capacity and operational optimization of both resources and pumping facilities.
Adjusting pump schedules.	Optimize pumping regime to match supply and demand, minimize energy requirements and retention times (linked to optimizing system storage).
Taking storage out of service or removing tank cells	Remove surplus capacity from system.
Reducing the TWL* of storage facilities.	Eliminate or reduce strategic storage and manage diurnal storage to minimize retention time. The TWL may be varied with seasons.
Altering storage configuration to avoid dead zones	Replace common inlet outlet with separate pipes or install baffles to eliminate dead water zones and improve throughput thereby reduce retention time.

* TWL = Top Water Level

CHAPTER 5
UTILITY SURVEY

INTRODUCTION

Management and maintenance of a distribution system involves many operational practices, some of which alter retention times. Often these practices are not used with the intention of improving water quality. However any change which reduces retention time will ultimately benefit water quality.

In the preceding chapters, various techniques for analyzing and resolving retention time problems were discussed. In order to understand what techniques are currently used or are viable for managing water quality, a questionnaire was circulated to a selection of international utilities who were known, or likely, to be involved in controlling water age. The results of and conclusions from the survey are reported in this chapter.

QUESTIONNAIRE

Details

A copy of the questionnaire is included in Appendix 1. The questionnaire was structured around the conclusions of Chapters 2 to 4 and had the following aims:

- To review the significance of various water quality problems and whether retention time has been used to control them
- To assess the feasibility of various methods for analyzing retention time problems
- To assess the feasibility of various methods for controlling retention time

The questionnaire comprised a summary discussion on retention time and the various tools for investigating and controlling water age, and 13 forms to gather information on:

Form 1: Company details
Forms 2 to 10: Specific water quality parameters influenced by retention time:
- (2) Loss of disinfection residual
- (3) High microbial counts
- (4) Animal infestation
- (5) Discolored water
- (6) DBPs
- (7) Polynuclear aromatic hydrocarbon (PAH) from bitumen lined pipes
- (8) Corrosion of metals
- (9) Taste and odor
- (10) Nitrification

Form 11: Any additional parameters not covered by Forms 2 to 10
Form 12: Assessment of feasibility of retention time management techniques
Form 13: Assessment of feasibility of retention time management analysis tools

The questionnaire was sent to all participating utilities and a selection of international organizations that were believed to be involved in such work. Although copies of the survey were forwarded to a breadth of international experience, responses were received only from Australia, the Netherlands, the UK, and the USA. In all, 23 utilities completed the questionnaire.

Results

Form 1 – Company Details

Utilities ranged in size from 100,000 to 4.3 million service connections. As one would expect, the number of water quality parameters investigated by a utility was not related to the size of its customer base. The most common issues of concern to utilities were discolored water (74 %), loss of disinfection (74 %), microbial (61 %), and taste and odor (61 %). Approximately 30 to 40 % of the utilities had problems with DBPs, corrosion of metals, and nitrification. Only one of the utilities that responded had investigated leaching of PAHs from bitumen lined pipes. The responses from the 23 completed surveys are summarized in Table 5.1.

Form 12 – Feasibility of Retention Time Management Techniques

The objective of Form 12 was to ascertain those physical and operational management techniques that the water industry considered to be pragmatic and effective solutions for reducing retention time. Table 5.2 summarizes the answers to Form 12.

With the exception of manual flushing, none of the techniques used to control retention time have been used by the majority of utilities surveyed. Adding those that are "likely to use" a technique, only four of the techniques were considered realistic options by more than 50 % of the respondents (manual flushing, adjust pump schedules, reducing the operational TWL, and altering storage configuration to remove dead zones). Conversely, of the options, time variable valves and downsizing pipes were not considered likely to be used by about 50 % of the respondents. However four of the five Dutch utilities have already downsized some pipes in their network, although the fifth utility declared that they would not use the technique.

While nine of the respondents had removed storage capacity (cells or whole tanks) from their systems, nine utilities did not believe they were likely to consider it an option. However the overall responses to the three storage techniques indicate that reducing storage and eliminating dead zones within storage are collectively realistic options.

Form 13 – Feasibility of Retention Time Management Analysis Tools

Form 13 addresses the viability of different analysis tools and whether they have been used to assess retention time. Of the 23 responses to the survey, 19 utilities completed this form. The results are summarized in Table 5.3.

The questions differentiated between whether the technique was viable and if it had been used. Generally those that had used a particular technique deemed it a viable option. However three of the 19 respondents deemed that manual calculations were not viable although they had used them; reasons given included "time intensive" and "not very reliable".

For each technique, at least half the utilities indicated that it represented a viable tool. However, only two utilities had used neural network models and optimization software. It is also of note that CFD, water quality statistical modeling, and physical models have not been used extensively. All respondents consider water quality data and using hydraulic models to be viable and are already using the techniques or are planning to do so.

Table 5.1
Parameters of concern to the surveyed companies

Parameter	Proportion of the 23 companies for which the parameter is an issue of concern	
	Number of responses	Percentage of responses
Discolored water	17	74%
Loss of disinfection residual	17	74%
High microbial count	14	61%
Taste and odor	14	61%
Corrosion of metals	10	43%
Nitrification	9	39%
Disinfection by-products (DBPs)	8	35%
Animal infestation	4	17%
PAH leaching from bitumen lined pipes	1	4%
Other parameters:		
Turbidity	1	4%
pH	1	4%
NDMA*	1	4%

* NDMA = N-nitrosodimethylamine

Table 5.2
Feasibility of retention time management techniques

Technique	Likelihood that companies will use the technique (number of utilities)					No response
	Will not use	Unlikely	Possible	Likely	Already in use	
Altering valving in network	1	2	8	3	8	1
Install time variable valve	3	8	8	1	2	1
Manual flushing	1	2	1	2	16	1
Automatic flushing	3	3	10	6	0	1
Abandon mains	3	4	6	3	5	2
Downsize pipes	3	9	2	1	7	1
Adjust pump schedules	0	2	6	7	7	1
Reduce the operational TWL	2	4	4	3	9	1
Remove reservoir cells	4	5	4	0	9	1
Alter reservoir configuration to remove dead zones	2	2	6	2	10	1

Table 5.3
Feasibility of retention time management analysis tools

Analysis tool	Percentage of the 23 utilities which consider the analysis tool to be viable	
	Viable	Used
Review water quality data	100%	100%
Hydraulic network model	100%	95%
Rules of thumb or practical experience	95%	95%
Water quality network models	86%	38%
Manual calculations	81%	90%
Customer perception surveys	81%	67%
GIS	81%	48%
Taste tests	76%	71%
Tracer studies	71%	57%
Reservoir model	67%	48%
Optimization software	62%	10%
CFD models	62%	33%
Physical model	52%	38%
Water quality statistical model	52%	14%
Neural network models	48%	10%

Form 2 – Loss of Disinfection Residual

Six of the companies surveyed did not cite loss of disinfection residual as an issue for them. Five of these companies were Dutch and do not practice residual disinfection. However, it was an issue for all of the other companies. The problem was generally identified by regular sampling supplemented by specific investigations. The cause was predominantly deemed to be system retention time and the majority of solutions involved managing the system including reservoir storage. However, seven of the sixteen utilities increased free chlorine and chloramine dosing as part of their improvement strategy. Most companies were able to demonstrate improved water quality through sampling.

Many companies used several improvement techniques. Reducing the TWL of storage or reconfiguring storage, predominantly by replacing common inlet and outlet arrangements with separate inlet and outlets, was used as an important technique by 70 % of the companies. This was followed by manual flushing (52 %), boundary valve changes (43 %), and rescheduling pumping regimes (35 %).

Few utilities used hydraulic or other modeling techniques to evaluate the problem. The majority analyzed and reviewed water quality data (61 % of companies), rules of thumb (52 %), and manual calculations (43 %) when formulating solutions.

Form 3 – High Microbial Counts

Thirteen utilities completed this form. These companies had identified high microbial counts (HPCs, coliforms, and *Aeromonads*) primarily through regular sampling. However, one utility stated that the problem had been identified through a specific monitoring program. Three of the utilities had not addressed the problem by controlling retention time.

There was a range of causes cited originating at the treatment works and within distribution. Retention time was identified as one of the causes. The main method of analyzing the problem involved reviewed water quality data (39 %). Other techniques were used including rules of thumb, manual calculations, and modeling. One utility used CFD modeling.

Various improvement techniques were adopted including reconfiguring and reducing storage capacity (35 % of companies), manual flushing, abandoning mains, boundary valve changes, and rescheduling pumping regimes. Generally data had been collected which demonstrated improved water quality.

Form 4 – Animal Infestation

Four utilities reported animal infestation, of which only two related the problem to retention time. The utilities resolved the problem through a combination of flushing and system reconfiguration. Sampling demonstrated that the problem had been resolved.

Form 5 – Discolored Water

Seventeen utilities reported that they had discoloration problems within distribution. Ten responses related their problems to cast-iron pipes and five to system operations issues. Eight utilities stated that some or all of their problems related to treatment processes. In addition, five utilities did not link managing discoloration to controlling retention time.

Eleven utilities adopted flushing to control their problem. A number of additional techniques were also implemented by individual utilities. Methods of analyzing the problem were generally limited to reviewing quality data, hydraulic modeling, and rules of thumb.

The effectiveness of the solutions was demonstrated through water quality data for eight of the twelve utilities that related the problem to water age. Seven other utilities confirmed that they perceived that water quality had improved and one utility did not consider that there had been a change in quality.

Form 6 – DBPs

Seven utilities reported THMs, HAAs, or bromate to be water quality problems within distribution. Three replies associated retention time, in storage or in the system with the problem, and five utilities identified the cause or part of it to the source water or treatment process.

One utility is currently using CFD modeling to analyze individual tank performance.

Three utilities used techniques that control retention time to manage water quality including rezoning and reducing storage operating levels. Two of these utilities demonstrated water quality improvement through sampling data, the other is yet to commission the works. Four utilities did not consider controlling retention when addressing the problem. Reasons for not using retention time control included a lack of confidence that it would solve the problem and cost implications.

Form 7 – Leaching of PAHs from Bitumen Lined Pipes

One utility's survey reported that PAH represented a water quality problem within their systems. The problem was rectified by controlling retention time using manual flushing and abandoning parts of the mains. Water quality improvement was demonstrated through sampling data.

Form 8 – Corrosion of Metals

Ten utilities responded that corrosion of metals was an issue within their systems. One utility related the cause to source water quality and provided additional treatment to resolve their problem. Two utilities identified the cause as unlined cast iron pipes, while seven utilities linked the cause to galvanized iron, copper, and lead pipes mainly within customer service pipes.

Five utilities identified that controlling retention time improved water quality including rezoning and reducing storage operating levels. However five utilities did not consider controlling retention time when addressing the problem. Reasons for not using retention time control included 'not confident that it would solve the problem', 'cost implications', and 'not identified as a controlling factor.

Two of the utilities that controlled retention time as part of their solution demonstrated water quality improvement through sampling data.

Form 9 – Taste and Odor

Of the 14 utilities reporting taste and odor problems, nine of them associated the cause with source water quality and resolved their problem by treatment and switching sources. Nine utilities identified retention time as a contributing factor. Two utilities are currently using CFD modeling to analyze storage performance.

Engineering and operational changes to control retention time were divided between network improvements; valve changes, manual flushing, and the operating regime; pump scheduling; and storage management.

Eleven utilities demonstrated water quality improvement through sampling data or perceived that there had been an improvement.

Form 10 – Nitrification

Eight of the nine utilities reporting nitrification problems confirmed that they had addressed the problem by controlling retention time. The other utility recognized that water age was a cause along with seasonal water temperature, which they resolved by secondary disinfection.

Three utilities used network and reservoir modeling techniques to analyze the problem. Seven utilities used rules of thumb and seven utilities reviewed quality data to analyze the problem.

The majority of utilities adopted manual flushing to improve water quality. However operational changes including valve changes and pump scheduling were employed as well as physical and operational changes to storage facilities.

All the utilities that adopted controlling retention time demonstrated water quality improvement through sampling data.

Form 11 – Other Parameters

Turbidity: One utility reported turbidity greater than 0.5 ntu caused by iron corrosion and rust sediments to be a particular problem. The utility used flushing to improve water turnover and hence water quality.

pH: Although a second utility did not consider controlling retention time to manage variable pH in distribution, they recognized that the cause of the change was related to cement lined pipes and age of water. Water quality was improved by flushing the systems and downsizing pipes.

N-nitrosodimethylamine (NDMA): One utility reported the presence of NDMA which it attributed to water age. There was a perceived improvement in water quality after adjusting pump schedules and reducing the TWL of storage.

Discussion

Viability of Tracer Studies

Tracer studies were considered to be viable by 71 % of the utilities surveyed. Overall 57 % had used tracer studies as an analysis tool. The use of the following trace chemicals was cited:

- Fluoride – Used by 6 utilities
- Calcium chloride – Used by 1 utility
- Lithium chloride – Used by 1 utility
- Sodium chloride – Used by 2 utilities
- Chlorine – Used by 1 utility

Where chlorine was used for the trace chemical, a spike concentration greater than the normal dose concentration was injected to trace its progress with time.

Methods Used to Control Quality Problems

Table 5.4 summarizes the water quality problems experienced by the utilities surveyed. The responses where retention time was identified as significant for the solution have been separated by parameter and source of the problem. It is worth noting that for five parameters, a utility did not resolve the problem and for two parameters, the cause of the problem was not known, although for one case of high microbial count, investigations were continuing.

Table 5.4
Methods used to control water quality problems (23 Utilities)

Water quality parameter	Problem investigated by utility	Did solution involve retention time management?		Solution involved adjustment of			Cause of problem unknown	Problem not rectified as still being analyzed
		Yes	No	Source or treatment only	Network problem Only	Source (or treatment) & network		
Disinfection residual	17	16	1	2	7	6	0	2
High microbial count	14	10	4	4	3	4	2	4
Animal infestation	4	2	2	0	2	2	0	0
Discolored water	17	12	5	0	9	8	1	1
Disinfection by-products	8†	3	4	5	1	1	0	0
PAH leaching	1	1	0	0	1	0	0	0
Corrosion of metals	10	5	5	3	3	2	0	1
Taste and odor	14	9	5	8	4	2	0	0
Nitrification	9	8	1	1	3	4	0	1*
Other parameters								
Turbidity	1	1	0	0	1	0	0	0
pH	1	1	0	0	0	1	0	0
NDMA	1	0	1	0	0	0	0	0

* Problem only partly solved
† Incomplete response from one utility

Retention Management Techniques

Table 5.5 summarizes the retention management techniques which have been used. For this table, the water quality parameter, "turbidity" which has been separately identified by one utility, has been included in the data for discolored water. The table notes where a technique has been used to control a parameter problem and, separately, quantifies the number of utilities that recognized that retention time was an issue as against those utilities that have used a retention time control technique, but not necessarily related the problem to retention time.

Mains flushing constitutes the main method of controlling retention time, it being used to manage all nine parameters. All of the techniques have been used for a minimum of five water quality parameters except time control valves and auto flushing, which have been used to control only two parameters.

It is of note that downsizing mains has not been considered for managing DBPs and taste and odor. However, adjusting pump schedules has been used for the majority of quality issues.

Analysis Tools

Table 5.6 summarizes the use of the different analysis tools for the range of water quality problems. The upper part of the table identifies where a tool has been used to assess a particular problem and the number of utilities that have used it.

Although all utilities consider reviewing water quality data and hydraulic modeling to be viable tools, nearly twice as many utilities actually use the former approach to analyze these problems compared with modeling.

It is of interest to note also the "rules of thumb" are deemed more viable and are used more than all other tools except for reviewing water quality data and hydraulic modeling.

There has been relatively low use of water quality statistical analysis techniques, storage, physical and CFD models, and optimization software. However approximately 60 % of the surveyed utilities considered the tools viable. Although a large number of utilities have used manual calculations to evaluate retention time (84 %), only 79 % considered them to be a viable analysis tool. This is an interesting contrast compared with the effectiveness of "rules of thumb".

Although GIS and water quality models have not been widely used, they are perceived as key potentially useful tools by over 80 % of the utilities.

Table 5.5
Usage of retention time management techniques

Water quality parameter	Alter system valves	Time control valve	Manual flushing	Auto flushing	Abandon mains	Downsize pipes	Adjust pump schedules	Reduce reservoir TWL	Remove cells & reservoirs	Reconfigure storage & network	Other techniques used
Disinfection residual	✓ (10)*	✓ (3)	✓ (11+1)	✗	✓ (3)	✓ (3)	✓ (8)	✓ (10)	✓ (4)	✓ (6)	- Fill and draw policy - Continuous flush - Mains extensions
High microbial count	✓ (6)	✗	✓ (7+1)	✗	✓ (6)	✓ (3)	✓ (3)	✓ (4)	✓ (4)	✓ (4)	- Strict sampling procedures
Animal infestation	✓ (1)	✗	✓ (2)	✗	✓ (1)	✓ (1)	✗	✗	✗	✓ (1)	
Discolored water and turbidity	✓ (3)	✗	✓ (11)	✓ (1)	✓ (2)	✓ (3)	✓ (1)	✗	✓ (1)	✗	- Bleeds & blowoffs (2) - Corrosion inhibitors - Pipe rehabilitation (2)
Disinfection by-products	✓ (2)	✗	✓ (1)	✗	✗	✗	✓ (1)	✓ (2)	✓ (1)	✓ (2)	- Continuous flush - Mains extensions
PAH leaching	✗	✗	✓ (1)	✗	✓ (1)	✗	✗	✗	✗	✗	
Corrosion of metals	✓ (1)	✗	✓ (4)	✓ (1)	✓ (1)	✓ (2)	✓ (1)	✗	✗	✗	
Taste and odor	✓ (4)	✗	✓ (5)	✗	✓ (2)	✗	✓ (4)	✓ (2)	✓ (1)	✓ (2)	- System reconfiguration - Change storage ops
Nitrification	✓ (5)	✓ (1)	✓ (5)	✗	✓ (1)	✓ (1)	✓ (5)	✓ (4)	✓ (4)	✓ (5)	
pH†	✗	✗	✓ (1)	✗	✗	✓ (1)	✗	✗	✗	✗	
NDMA	✗	✗	✗	✗	✗	✗	✓ (1)	✓ (1)	✗	✗	

* The number in brackets indicates the number of utilities which have used the technique, (9+1) refers to nine utilities controlling retention time and one utility that adopted a technique that is effectively controlling retention time

† Additional parameter investigated by one utility

Table 5.6
Usage of retention time management analysis tools

Water quality parameter	Manual calc-ulations	Rules of thumb	Review water quality data	GIS	Hydraulic models	Water quality models	Water quality statistics	Reservoir models	Physical models	CFD models	Taste tests	Customer perception	Optim-ization software
Disinfection residual	✓ (10)*	✓ (12)	✓ (14)	✓ (4)	✓ (6)	✓ (3)	✓ (1)	✓ (2)	✓ (1)	✓ (4)	✗	✗	✓ (1)
High microbial count	✓ (6)	✓ (6)	✓ (9)	✓ (4)	✓ (5)	✓ (2)	✓ (1)	✓ (2)	✓ (1)	✓ (2)	✗	✗	✓ (1)
Animal infestation	✓ (2)	✓ (1)	✓ (2)	✓ (1)	✓ (2)	✓ (1)	✗	✗	✗	✗	✗	✗	✗
Discolored water and turbidity	✓ (2)	✓ (7)	✓ (9)	✓ (2)	✓ (5)	✓ (1)	✓ (1)	✗	✗	✗	✗	✗	✗
DBPs	✓ (2)	✓ (2)	✓ (3)	✗	✓ (2)	✗	✗	✗	✗	✓ (1)	✗	✗	✗
PAH leaching	✓ (2)	✓ (1)	✓ (1)	✓ (1)	✓ (1)	✗	✗	✗	✗	✗	✓ (1)	✗	✗
Corrosion of metals	✓ (2)	✓ (3)	✓ (4)	✓ (2)	✓ (2)	✓ (1)	✓ (1)	✗	✓ (2)	✗	✗	✗	✗
Taste and odor	✓ (2)	✓ (6)	✓ (4)	✓ (2)	✓ (1)	✓ (1)	✗	✓ (1)	✗	✓ (2)	✓ (4)	✓ (2)	✗
Nitrification	✓ (2)	✓ (7)	✓ (7)	✓ (1)	✓ (4)	✓ (2)	✓ (2)	✓ (2)	✗	✓ (1)	✗	✗	✗
pH	✓ (1)	✓ (1)	✓ (1)	✗	✗	✗	✗	✗	✗	✗	✗	✗	✗
% of utilities who consider the tool viable	✓ (1)	✗	✓ (1)	✗	✗	✗	✗	✗	✗	✗	✗	✗	✗
% of utilities who have used the tool	81%	95%	100%	81%	100%	86%	52%	67%	52%	62%	76%	81%	62%

* Number in brackets is the number of utilities who have used the analysis tool to investigate the problem

CHAPTER 6
CASE STUDIES AND WORKSHOP

WORKSHOP OVERVIEW

A workshop was held in November 2002 attended by representatives from utilities, academics, and industry consultants. The aims of the workshop were:

- To disseminate and gain feedback on the initial findings of the project
- To gather information from utilities on their experience of using retention time management
- To discuss and identify the knowledge gap, the needs of utilities, and how they should be addressed in Phase II of the project

Based on the results from the questionnaires (Chapter 5) a selection of case studies was identified. Invited utilities gave presentations on a range of potential retention time related water quality issues and selected topics of interest. Some of the presentations gave details of various analysis and investigation tools which were relevant for investigating retention time related water quality problems. Details from these have been incorporated into Chapter 3. Other utilities presented case studies of water quality problems they have addressed where controlling retention time was part of the solution. These case studies are summarized in this chapter.

CASE STUDIES

Case Study 1: Simple Manual Calculation of Retention Time

This case study was provided by Thames Water Utilities, UK. It illustrates the use of simple manual calculations to perform quick checks. This case study uses the methods and formulae presented in Chapter 3 in the section *Manual Calculations*.

Problem

A secondary pumping station, boosting water from a neighboring distribution zone provides an important peak and emergency backup function to the main groundwater source supplying a major urban area. Due to water quality concerns, the utility wanted to ensure that the water in the main was turned over every two days. The utility performed the following simple calculation to estimate the minimum pump run time during off-peak periods required to achieve this.

Analysis

Length of pumping main 'L' = 4 km = 4000 m
Diameter of pumping main 'd' = 600 mm = 0.6 m
Pump output flow 'Q' = 150 L/s = 0.15 m^3/s

hence

$$\text{Pipe area } A = \frac{\pi d^2}{4} = \frac{\pi 0.6^2}{4} = 0.283 \text{ m}^2$$

$$\text{Output velocity } u = \frac{Q}{A} = \frac{0.15}{0.283} = 0.530 \text{ m/s}$$

$$\text{Retention time } t = \frac{L}{u} = \frac{4000}{0.530} = 7550 \text{ s} = 2.1 \text{ h}$$

Solution

Hence to ensure a turnover of water in the pipeline every two days the pumps have been scheduled to run for one hour every day during periods of low electricity tariff. In addition the output velocity of 0.53 m/s is probably sufficient to ensure that the pipeline is swept regularly of settled material that may otherwise build up and aggravate water quality conditions. A minimum pipe velocity of 0.4 m/s would generally be accepted as reasonable to prevent the build up of settled material.

Key Points

- For simple pipelines, retention time can be calculated using simple manual calculations
- The target maximum retention time of two days was based on the experience of the utility. Maximum acceptable retention times will vary considerably depending on water chemistry and pipe conditions.

Case Study 2: Spreadsheet Model to Calculate Retention Time

This case study was also provided by Thames Water Utilities, UK. It illustrates how spreadsheet models can be used to calculate retention time in simple systems with time varying flow patterns. This case study uses the methods presented in Chapter 3 in the section *Spreadsheet Models*.

Problem

This case study is of the system shown in Figure 6.1. A 15.8 ML tank is fed by a pumping main approximately 10 km long. A further 4 km of mains link the tank to the town (population 34,000). Originally the tank was connected by a common inlet and outlet main at the site and suffered from occasional coliform failures particularly during the summer months.

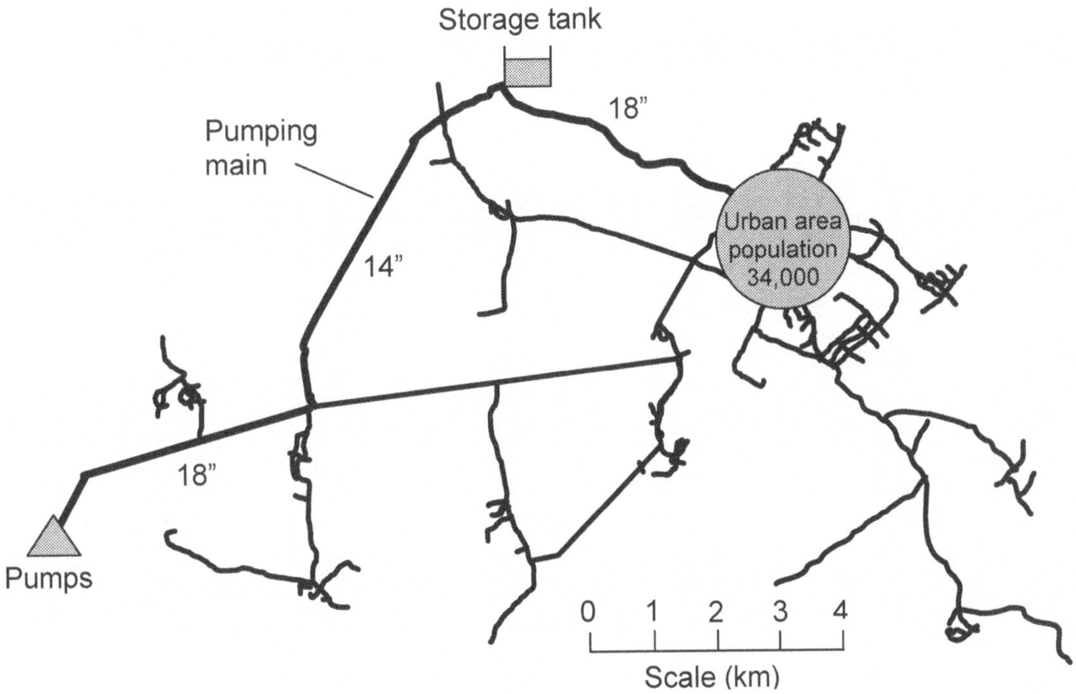

Figure 6.1 Case study 2 – network plan

Analysis

A spreadsheet analysis of the tank turnover based on the water level profile from telemetry records (Figure 6.2) indicated that the turnover was particularly poor, less than 50 % of the water in the reservoir was replaced every 5 days, even based on an assumption of perfect mixing within the reservoir.

Using the recorded flows in the pumping main it was also possible to estimate by how much the turnover could be improved by creating dedicated inlet and outlet mains at the storage tank site. Figure 6.3 indicates a substantial improvement.

Solution

On the basis of the above analysis the reservoir was modified so that it had a separate inlet and outlet and further coliform failures were prevented.

Key Points

- The data required for the analysis was readily available from telemetry
- Simple analysis and modifications to the reservoir rectified a potentially serious problem with noncompliance of water quality regulations
- The spreadsheet uses similar formulae to those which would be used for manual calculations, but reduces the effort in accounting for time varying flow patterns

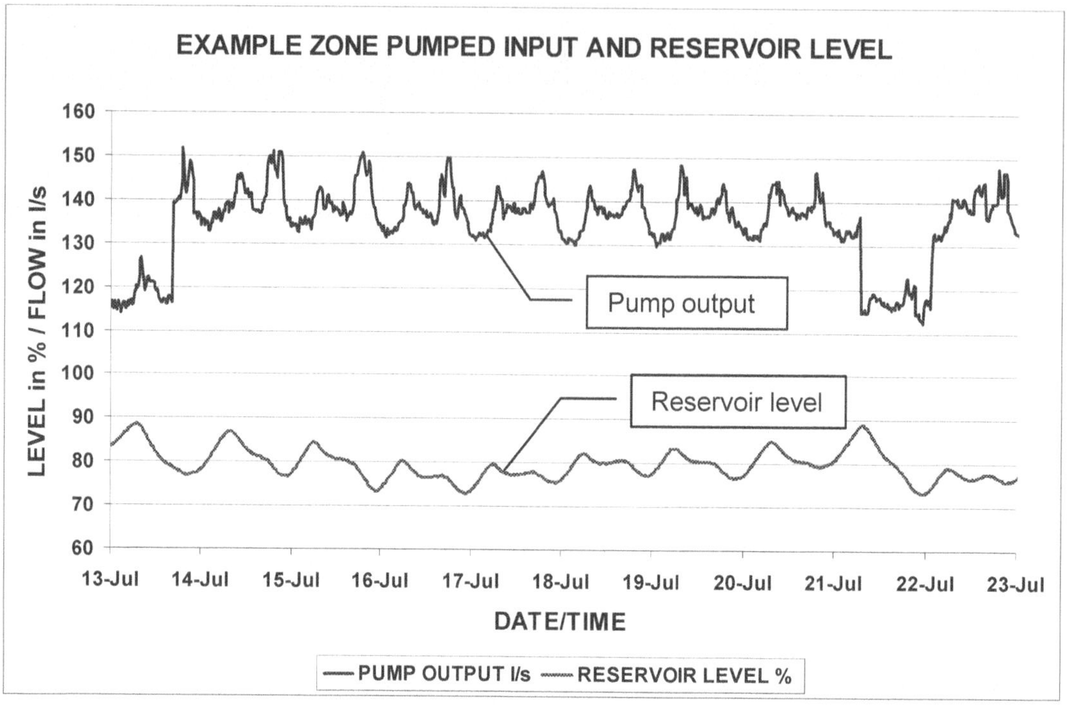

Figure 6.2 Case study 2 – input data from SCADA system

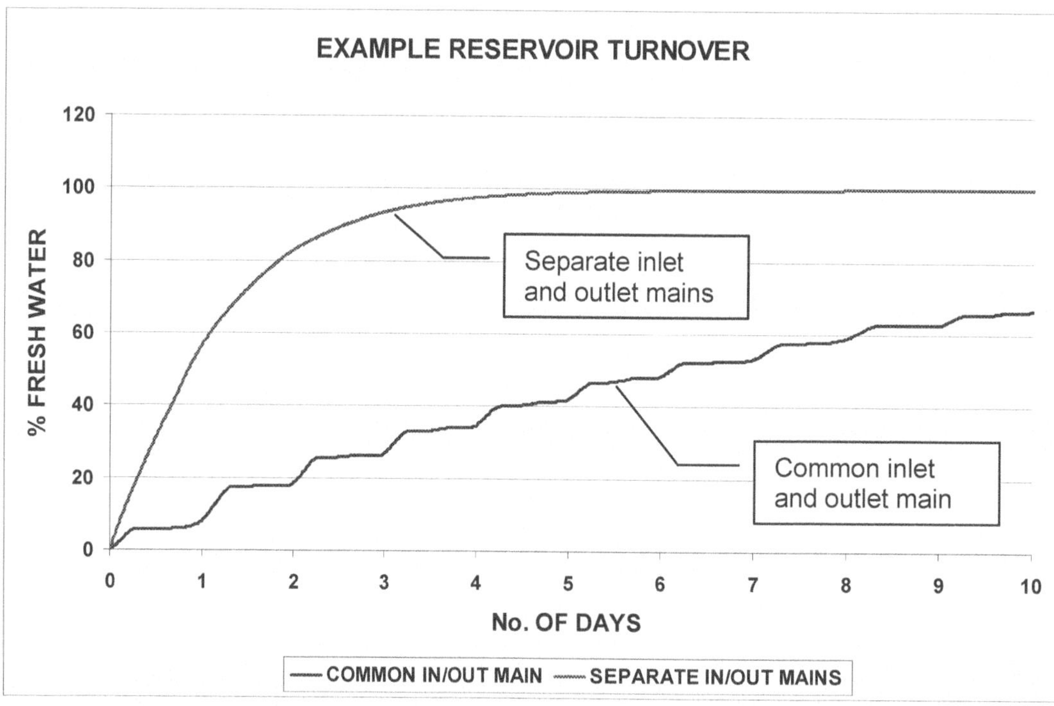

Figure 6.3 Case study 2 – spreadsheet estimates of reservoir turnover

Case Study 3: Pipeline and Booster Improvements

This case study was presented by Thames Water Utilities, UK. The utility experienced problems of poor microbial water quality in the water supply to a town and surrounding rural area with a total population of approximately 10,000. The case study gives details of the methods used to investigate the problem and assess a range of possible solutions.

Problem

Storage Tank B was constructed to provide security of supply to the town previously supplied only by Tank A (Figure 6.4). Tank B supplies the zone via a single 450 mm common inlet and outlet main. A remotely actuated valve was installed on the 15-in. main supplying the zone to control the flow from Tank A, and hence control the level of Tank B. The network configuration, whereby water was fed to Tank B from Tank A via the 15-in. main and through the zone, made it difficult to turn Tank B over and lead to water quality problems with excessive coliform levels.

The company introduced some quick remedial measures about five years ago. These measures included the infusion of water from another zone and removing half the storage at Tank B from service. These actions maintained coliform levels within acceptable limits however, more recently there has been a need to reduce the infusion flow from the neighboring zone and so an alternative solution was required. It was also desirable to reinstate the full storage at Tank B.

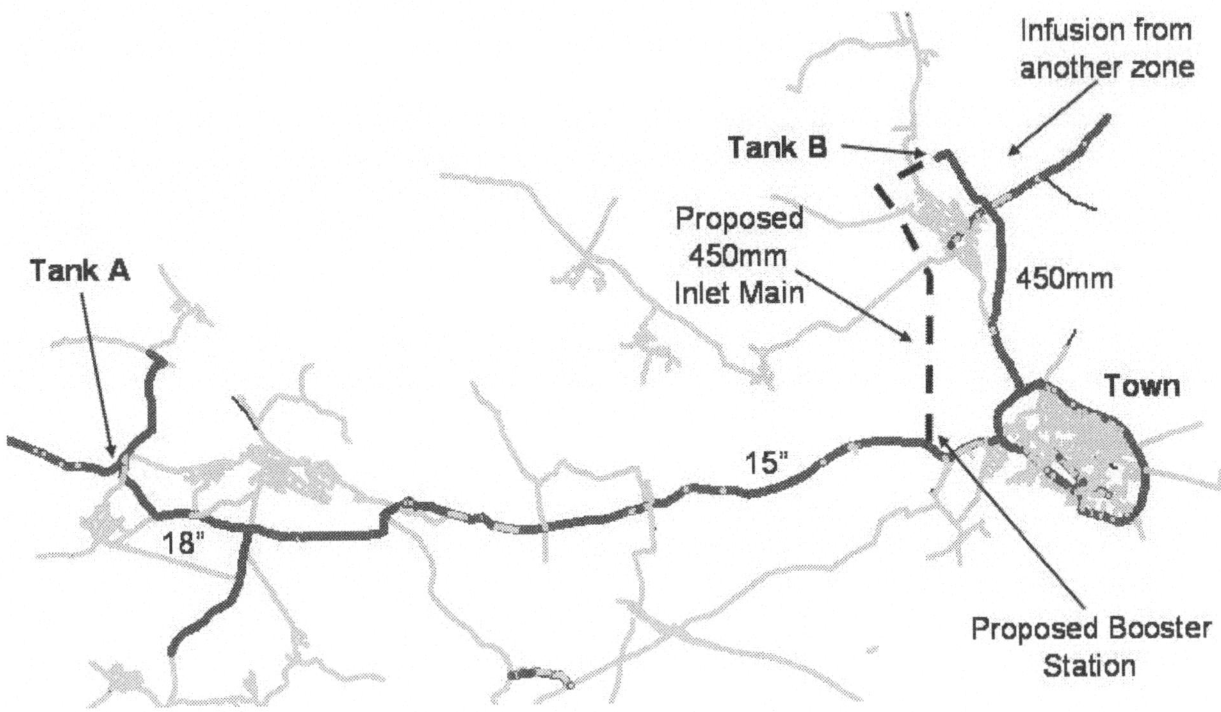

Figure 6.4 Case study 3 – network plan

Analysis

The analysis was primarily undertaken using an EPANET hydraulic and water quality network model. The model was updated with current demand data and used to analyze the hydraulic behavior, age of water, and chlorine decay under the current operational regime and three proposed options:

- *Option 1:* Duplicate the existing 450 mm main to create dedicated inlet and outlet mains and install a booster to pump water along the new main directly into the reservoir
- *Option 2:* Install only the booster
- *Option 3:* Install only the 450 mm main

The EPANET model predicts that the current configuration turns over approximately 50 % of Tank B in 5 days. While this has not so far caused water quality failures, this low rate of turnover gave some cause for concern. The model also demonstrated that the chlorine residual within the reservoir was entirely dependant on the infusion flow.

The model predicted that option 1 would significantly improve the turnover in Tank B to 75 % in 2 days even with the full storage reinstated. However, the existing 450 mm main, which would become the outlet main feeding the town, was known to be in poor condition and the chlorine decay rate in it was likely to be high. Consequently the chlorine residual levels in the town could become unacceptable. To rectify this the main may need to be rehabilitated or booster chlorination installed which would be expensive options. The model also showed that when the pump was in operation, there could be problems with reduced pressures elsewhere in the network.

The model predicted that Option 2 would give a similar rate of turnover in Tank B to the current setup. However, a substantial improvement could be obtained (60 % turnover in 2 days) by maintaining a small infusion from the neighboring zone at night while the tank filled and not reinstating the full storage. A number of valve adjustments were tried, but none represented any significant benefit. The model was also used to assess the size of pump that would be required for this option to be successful.

Hydraulic modeling showed that installing the 450 mm main without a pumping station (Option 3), the infusion flow will still be required to meet the demand in the town during peak periods. Hence this option was discounted.

Solution

Option 2 has been selected as this offers the potential of a two stage approach, proceeding to Option 1, if problems persist.

Key Points

- There was poor microbial water quality associated with inadequate turnover of storage
- Although a short-term solution had been implemented, there was the need for a long-term solution which addressed both supply and water quality issues

- The use of an EPANET model allowed the hydraulic behavior, water age, and chlorine residuals to be predicted and compared for several alternative solutions
- The model highlighted hydraulic limitations with some of the proposed options which may not otherwise have been apparent
- The analysis has shown that a significantly cheaper solution than that first considered may be adequate and may be preferable with respect to maintaining chlorine residuals in the town

Case Study 4: *Aeromonads* Problem

This case study was presented by Waterlaboratorium Noord, Netherlands. This case study gives details of microbial problems due to high retention time in a trunk main supplying a small city.

The *Aeromonads* species found in water are an environmental strain and are nonpathogenic. *Aeromonads* can be easily grown in low concentrations of nutrients, their response is quicker than a general colony count, and a low count indicates lower change of higher organisms. The species is used in the Netherlands as a parameter to monitor the biological water quality in distribution systems.

Problem

The city was supplied from a pumping station via twin trunk mains (1000 or 630 mm diameter PVC and 750 mm diameter cast-iron). In 1997 a third 800 mm diameter GRP pipe was laid between the pumping station and the water quality sampling in the two older pipes from the pumping station indicated high *Aeromonads* levels. *Aeromonads* levels in the new pipe and other trunk mains supplying the city remained low.

Analysis

A sampling program was set up to monitor *Aeromonads* and temperature in the trunk mains over a 6-month period. The program found a correlation between *Aeromonads*, temperature, and retention time (Figure 6.5). It was concluded that the problem had been caused by the significant increase in retention time which occurred when the new main was laid.

Solution

The proposed solution was to reline the old 750 mm iron pipe and significantly reduce its diameter. The relining would make the internal surface of the pipe less susceptible to microbial growth and reduce the retention time in the main and in the parallel 1000 or 630 mm diameter PVC main. It was believed that the reduction in retention time that would result should resolve the *Aeromonads* problems.

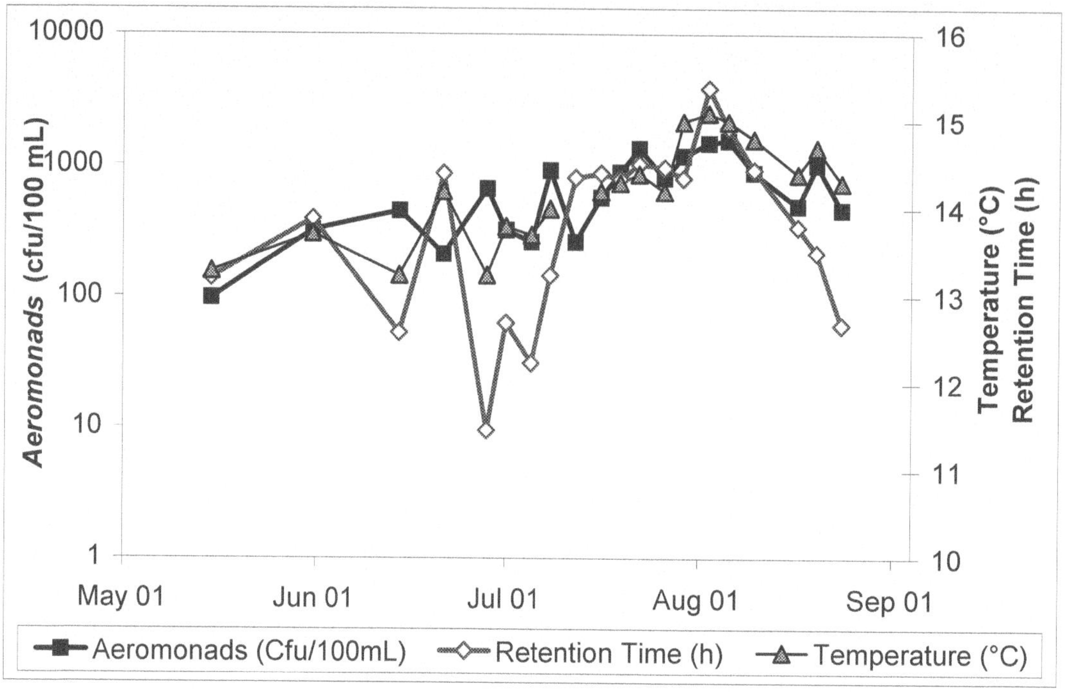

Figure 6.5 Case study 4 – *Aeromonads* retention time and temperature

Key Points

- A sampling program demonstrated a correlation between high retention time and high *Aeromonads* levels. It should be noted that in accordance with common practice in the Netherlands, the supply was not chlorinated. A similar correlation might not exist for chlorinated supplies.
- Increasing the hydraulic capacity of the network caused excessive retention times which resulted in poor microbial water quality.
- Relining was used as a means not only of improving the internal condition of a pipe but also of reducing its diameter and thereby reduce the retention time

Case Study 5: Improving Taste

This case study is based on a project which was undertaken for Severn Trent Water, UK (Brandt, Creasey, and Perkins 1999; Brandt 2000). The purpose of the project was to identify options for improving the consistency of chlorine residual received by 4.5 million customers.

Problem

An analysis of customer complaints and household survey data identified that some customers were unsatisfied due to a chlorine taste in their tap water. The Utility decided to improve the control of chlorine levels and its variability within distribution. Initial data analysis identified that there were approximately 270 potential problem areas in 180 chlorine control zones, serving about 4.5 million consumers. A study was undertaken with the aims of:

- Reducing ex-works chlorine residuals to a maximum of 0.4 mg/l
- Reducing the chlorine residual at the customer's tap to or below 0.3 mg/l
- Reducing variations in chlorine residual at any point to less than or equal to 0.2 mg/l
- Maintaining a minimum chlorine residual of 0.05 mg/l
- Not to compromising bacteriological compliance

Analysis

A wide range of data was collected and compiled for use in the various stages of the analysis process. For each of the chlorine control zones, this information included the record drawings of the distribution network, details of mains materials and condition, details of all rehabilitation and renovation in the previous three years, and an understanding of the operation of the system. Engineering and operational records for the storage facilities, including internal layout and inlet and outlet details were also assembled. Hydraulic network models were available for the majority of the chlorine control zones. Water quality data was available from the records of fixed and random location sample data for the source water, permanent chlorine and bacteriological sample points, the sample results for the previous 12 months from the company's water quality information system. Other company information system data collected included records from at least the 12 previous months for system leakage flow monitoring and telemetry flows and pressures. Location data, customer complaints and inspectors report data, and the results of customer perception surveys were used for the prioritization process.

Additional specific information was collected when identified by initial analysis that its inclusion would enable a positive recommendation to be made or where there was no clear understanding of the causes of potential problems in the zone.

Within the constraints of availability and quality of data, the analysis for each zone followed the same sequence and produced standardized outputs and conclusions. This was achieved by using a decision tree approach and prescriptive procedures to ensure consistency of output, and auditable conclusions. The analysis tools used included:

- Calibrated hydraulic models, extended and modified to reflect current operational conditions and skeletal hydraulic models of zones not previously modeled
- Age of water models developed from the hydraulic models
- Chlorine water quality models validated against contemporary sets of water quality sample data for summer and winter
- Spreadsheet analysis to calculate time of travel and chlorine decay of trunk mains and similar linear systems upstream of a DMA (Figure 6.6)
- Microbiological statistical modeling of the relationship between failure probability and chlorine levels to assess future risk

The chlorine models were used to analyze current and test proposed chlorine dosing regimes and to assess the risk of future microbiological failure as a consequence of reducing chlorine dosing rates. Capital and operating costs were prepared for each scheme using standard unit costs and the costs for the 360 individual schemes were then rolled up into 102 hydraulic supply areas. Customer complaints and perception survey data from the company data bases was then used together with operational and physical data for the 102 supply areas to develop a priority ranking scheme for implementing the recommendations.

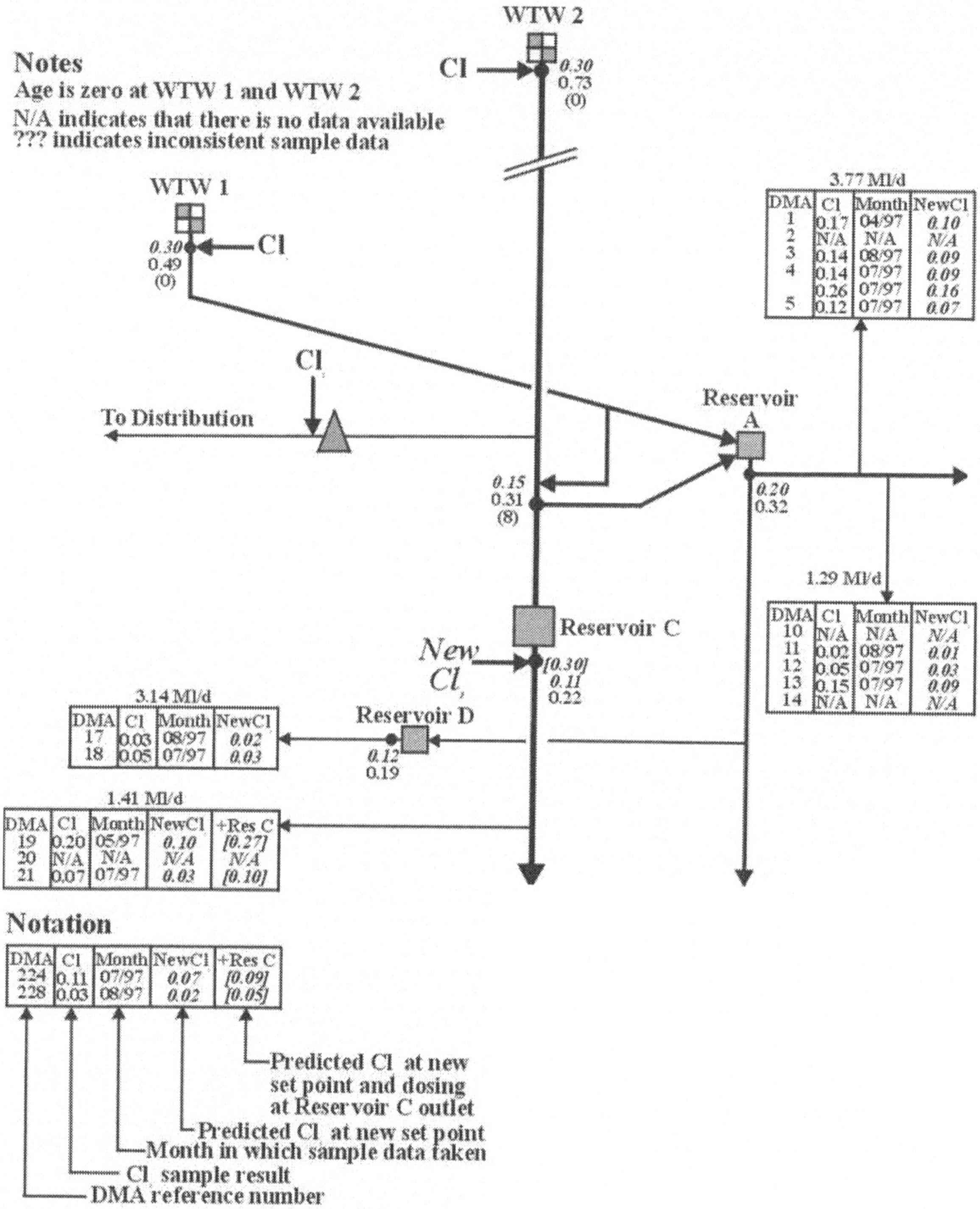

[0.17] Chlorine concentration mg/l with new set point and dosing at Reservoir C outlet

0.17 Chlorine concentration mg/l with new set point

0.22 Chlorine concentrations mg/l with existing set point

(12) Water age in hours (where determined)

Figure 6.6 Case study 5 – spreadsheet analysis of retention time

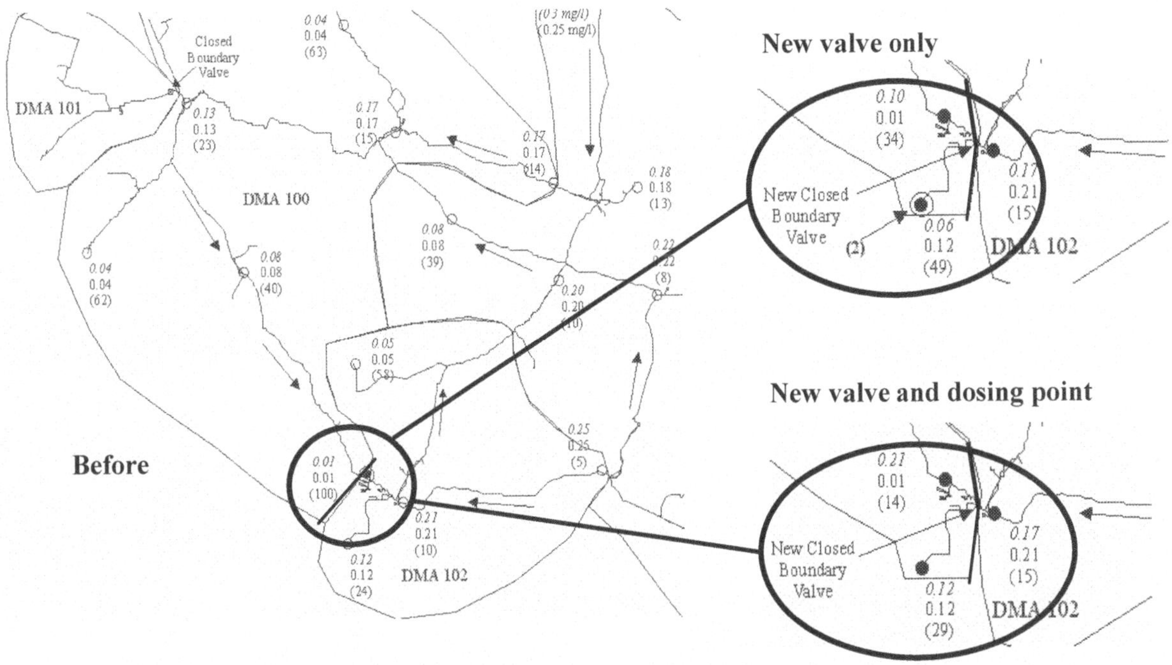

Figure 6.7 Case study 5 – rezoning to control time

Solutions

The majority of solutions involved changing dosing rates at existing facilities or introducing new injection points. However, other options to reduce chlorine decay rates and reduce water age were also considered including;

- Improving system hydraulics to increase velocities in pipes including rezoning and moving boundary valves (Figure 6.7)
- Rerouting supplies into an area
- Rehabilitating mains and storage facilities
- Reconfiguring flow paths to and within storage facilities
- Reducing or eliminating storage capacity (reducing top water levels and decommissioning storage cells)
- Storage tank, borehole, and pumping station operational changes
- For low risk zones, condition inspections at 15 sites identified 11 fault categories that were used to quantify "generic solutions" for all low risk zones
- Where there was no clear solution, further investigations were recommended

In practice, although an implementation program was recommended based on maximizing the number of customers to receive improved water quality, for business reasons the implementation program was incorporated into the operational maintenance programs.

Key Points

- A survey of the utilities' customers found chlorine taste to be an issue
- A large number of supply areas were analyzed using a standardized analysis approach involving hydraulic, age, and water quality analysis in conjunction with statistical biological compliance modeling
- The water quality analysis reviewed both summer and winter conditions
- A range of operational and engineering solutions were identified

Case Study 6: Aesthetic Water Quality

The Philadelphia Water Department presented details about their approach to investigating aesthetic problems arising from a number of related water quality problems.

Problems

The City has the following water quality problems in some areas of its network:

- Decay of chloramine residual – a 2.0-mg/l chloramine residual is provided in the water leaving the treatment works
- Corrosion of unlined cast-iron pipes
- Leaching of cement linings resulting in elevated pH
- Nitrification due to chloramine decay

Chloramine decay and nitrification are seasonal and are generally only a problem with the higher water temperature between April and October. The above water quality problems, although infrequent, can result in aesthetic problems which lead to customer complaints:

- Taste:
  - Earthy, musty, or stale flavors and odors are unmasked as the chloramine residual decays
  - Metallic tastes due to corrosion of iron mains and copper service lines
  - Astringent tastes due to leaching of cement linings and elevated pH
- Turbidity due to corrosion and deterioration of unlined cast-iron mains
- Discoloration and staining due to oxidation of ferrous to ferric iron

The causes of the problems include:

1. High water temperature in summer months – seasonal range of 5 °C to 28 °C
2. Unlined cast-iron pipes – over 50 % of mains in the city are unlined cast iron
3. Cement lined pipes – most new pipes are cement lined ductile iron
4. High retention times – diminished population and industry has resulted in a large surplus in network capacity including some oversized or underused mains
5. Closed valves – until the City completes a database of all its valves, and maintains that database with checking of valve positions, unknown closed valves create artificial dead ends throughout the City

Analysis

The problems were identified and investigated primarily by collecting and reviewing water quality data. In particular the City has found it beneficial to monitor and act on:

- *Chloramine levels*: Many of the problems are related to a loss of chloramine residual, hence by monitoring trends in chloramine levels it is possible to predict when these other problems are likely to occur. The City considers that total chloramine residuals below 0.5 mg/l often indicate water quality deterioration due to long retention times (based on initial dose of 2.0 mg/l).
- *Customer complaints about aesthetic quality*: This often indicates long retention times which are often caused by customer plumbing problems
- *Turbidity*: The City has found that samples with turbidity < 0.3 ntu generally have insignificant levels of iron whereas samples with turbidity > 2 ntu are likely to exceed the USEPA's guideline limit for iron. Above 1 ntu the City has observed a reasonable correlation between turbidity and iron (Figure 6.8).

Flushing through hydrants has helped to diagnose the cause of problems and can restore water quality for short times, but flushing has not been a long-term solution.

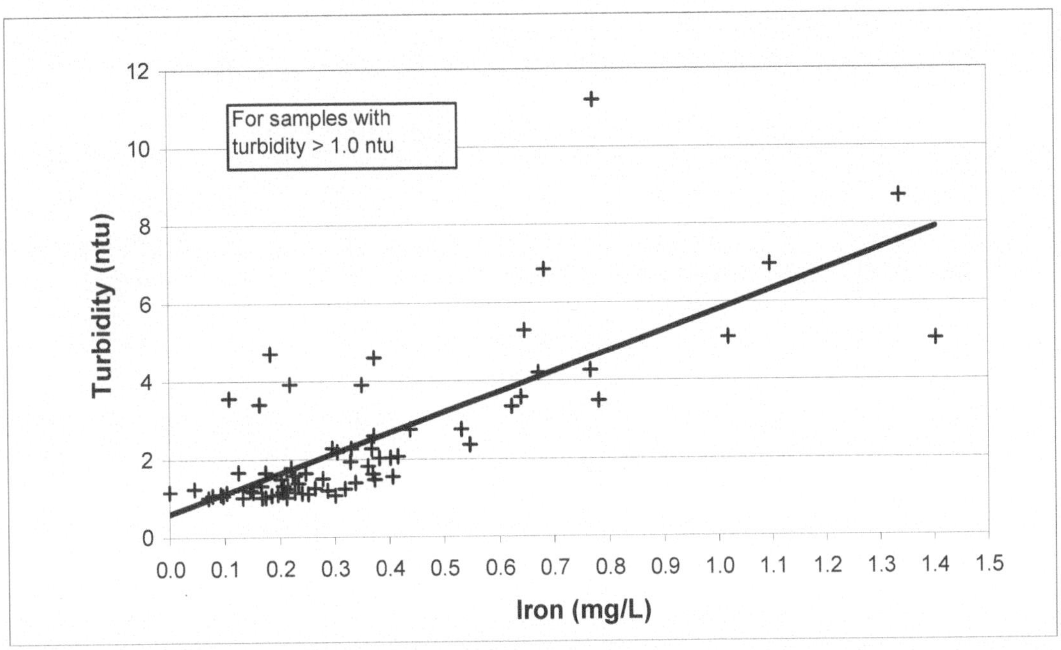

Figure 6.8 Case study 6 – correlation of turbidity v iron

Reprinted from Opflow, Vol. 28, No. 12 (December 2002), by permission. Copyright © 2002, American Water Works Association

Solutions

1. Optimize storage tank or reservoir operation to minimize retention time:
    - Reduce stored volume
    - Alter pump scheduling
    - Reconfigure from riding on system pressure to flow through mode operations
2. Replace or rehabilitate mains. This solution not only addresses the problems caused by unlined cast-iron pipes, but can also allow the internal diameter to be reduced or pipes to be reconfigured and thereby reduce retention time.
3. Keep track of valve operations and alter valve setup if appropriate. It is recognized within the department that long retention times can occur anywhere in a system when caused by closed valves. Hence, closed valves are always considered when investigating problems.

Key Points

- There can be a strong relationship between multiple parameters controlled by retention time. The Philadelphia Water Department has observed relationships between:
    - Chloramine residual
    - Nitrification
    - Taste
    - Discoloration
    - Turbidity
- The water quality problems showed strong seasonality
- Several different methods have been successfully implemented to resolve the problem by reducing retention time
- Their approach is driven by monitoring a few key water quality parameters

Case Study 7: Reengineering a Network to Reduce Retention Time

The Greater Cincinnati Water Works has experienced problems with insufficient chlorine residual in a new supply zone. The problem is caused by excessive retention times as the zone was designed and built to meet future demands, but currently there is little demand.

Problem

The network has been expanded into a new zone in which future development is planned. The new network has been sized to meet predicted future demand, but the initial demand in the area is low. Results from a number of sampling stations in the new network indicate insufficient chlorine residual.

Solutions

The company has implemented a number of operational control measures such as weekly flushing and altering tank operation. Although these measures generally produce immediate improvements, it is recognized that they have only short term benefits. A trial by the company found that at a number of sites flushing only gave improved chlorine residuals for a period of one day (Figure 6.9).

Reengineering solutions involve physical changes to the network such as mains construction or valve installation. It was understood that reengineering solutions were likely to be more expensive than operational solutions such as flushing, but would give permanent benefits and so be more effective in the longer term. Some examples of reengineering solutions are given below:

Division valve installation: By inserting shut valves it may be possible to lengthen the flow path to a point of high demand. While this will increase the age of water at that demand point, it will significantly reduce the retention time to all supply points along the new supply route. An example of this approach is illustrated in Figure 6.10. In this case the major demand is provided by an industrial consumer. By inserting a shut valve and forcing the flow around a loop to the south, it is possible to significantly improve water quality to existing customers fed off this loop.

Main extension and tie-in: By extending the network and then tying it back into the existing network (which has reasonable existing demands), turnover is achieved in the new mains across which there would otherwise be relatively little demand. An example of this approach is illustrated in Figure 6.11. This approach can significantly improve water quality in the new mains, but will increase the water age and reduce the water quality supplied to the parts of the existing network into which the new network is tied. Water quality network models can be valuable for predicting whether the water quality will be impaired excessively in such areas. Similarly, storage tanks in the new zone have been tied back into the existing network thereby preventing large volumes of water stagnating in the new storage tanks.

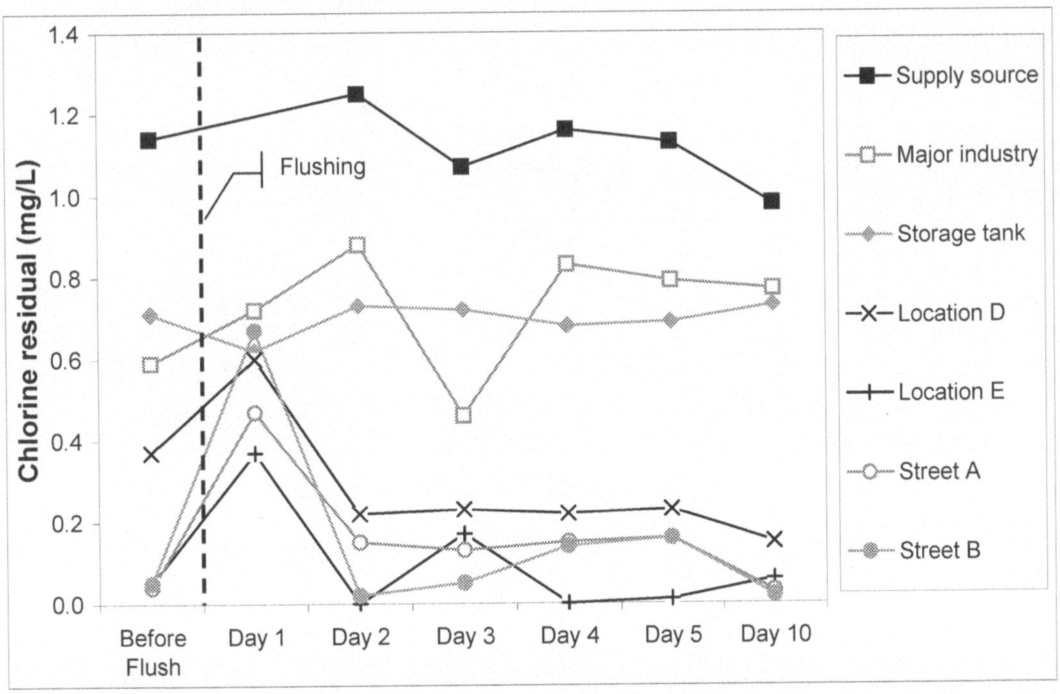

Figure 6.9 Case study 7 – chlorine residual following flushing

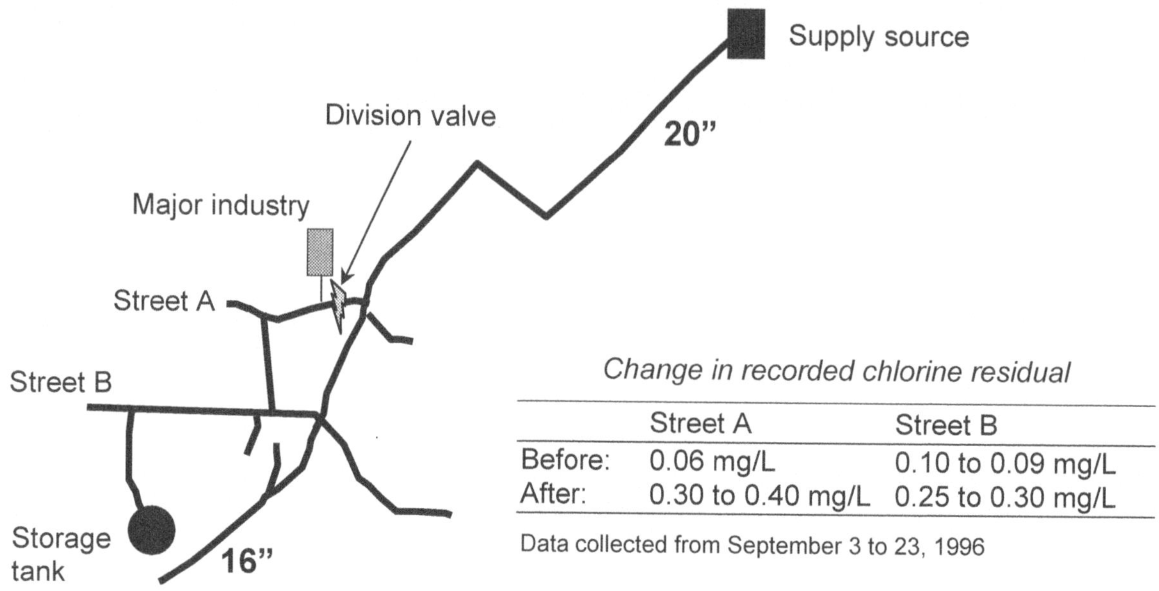

Change in recorded chlorine residual

	Street A	Street B
Before:	0.06 mg/L	0.10 to 0.09 mg/L
After:	0.30 to 0.40 mg/L	0.25 to 0.30 mg/L

Data collected from September 3 to 23, 1996

Figure 6.10 Case study 7 – division valve installation

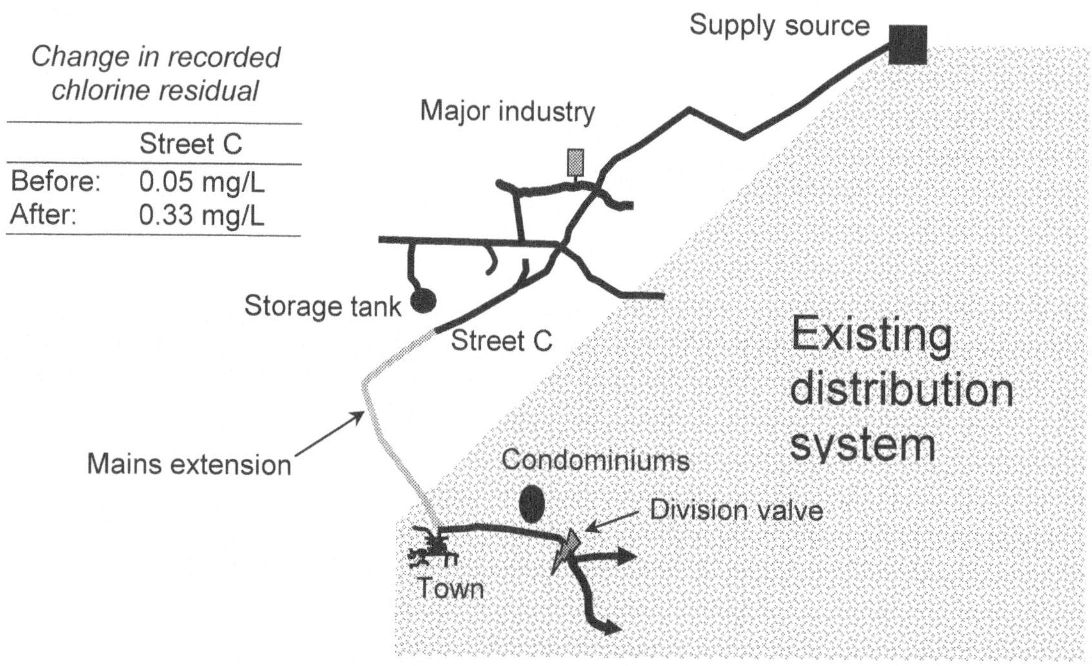

Change in recorded chlorine residual

	Street C
Before:	0.05 mg/L
After:	0.33 mg/L

Figure 6.11 Case study 7 – main extension and tie-in

Develop system to customer base: A third approach used was to look for potential existing demand centers and set up the network so that these demand centers are located on the hydraulic periphery of the new network. In this case study there were a number of mobile home parks which could be supplied in this manner.

Key Points

- Reengineering solutions are effective and give long-term benefit
- In this case flushing sometimes only gave benefit for one day
- Altering the valve setup of a network can improve water quality in some parts of the network, but will reduce it in other parts. Although there will always be a trade off between areas by any revalving solution, the aim should be to avoid large areas being fed by extremes in water age.
- Hydraulic and water quality network models are useful tools for predicting the potential benefits and disadvantages of reengineering solutions

Case Study 8: Storage Tank Inlet Reconfiguration

Champlain Water District has used CFD modeling to improve the design of water tanks in which poor turnover of water in tanks, exacerbated by thermal effects, causes problems with low chlorine residuals and DBP formation.

Problem

The company has a number of storage tanks in which the same pipe serves as both the inlet and the outlet. This results in 'standpipe' design behavior and their associated problems with poor turnover of water. The water temperature of the water supplied into the tank may vary diurnally from 7.5 °C to 15 °C or higher. Theoretically, the temperature at the top of the tank may rise to 25 °C. The density of water decreases with temperature and so the buoyancy of the water entering the tank may differ from that elsewhere in the body of the tank. Under some circumstances this could inhibit mixing within the tank and lead to stratification or cause older, less mixed water to be displaced from the tank. Past water quality investigations observed pockets of water with high DBPs and low chlorine residual in the tank and it was suspected that they may be due to stratification. The district wanted to develop a standard inlet and outlet design which would promote mixing and with minimal changes could be applied to all storage tanks which are similarly affected.

Solutions

CFD modeling of the existing tank design predicted that stratification would occur even when the temperature of the influent was only slightly (2 °F or 0.6 °C) less than the water in the main body of the tank. The model of a proposed alternative inlet located half way up the tank and angled at 45 °C from vertical indicated that significantly improved mixing could be achieved. In addition to this new inlet feature configured onto the existing pipe entering and exiting the tank, an outlet feature was added to withdraw water from the bottom of the tank, again using the same pipe that enters or exits the tank. The design has been implemented and is being evaluated with a complete survey to see and compare the relative rate of DBP formation and chlorine residual degradation before and after tank reconfiguration.

Key Points

- Thermal effects can inhibit mixing within tanks causing stratification and pockets of very old water
- This can result in variable water quality (DBP and chlorine residuals)
- CFD is useful for modeling and visualizing mixing within tanks and can be used to predict problems caused by thermal effects
- CFD indicated that a change in the inlet and outlet configuration may significantly improve mixing within the tank
- This configuration is currently being evaluated in a full-scale case study

Case Study 9: Monitoring Water Quality in Distribution

In Birkeroed, Denmark, a small region (200 properties) of the city's distribution network suffered from bad taste and discolored water at certain times of the year. Age of water modeling showed that this was due to a hydraulic balance point in the network.

Problem

The company received a number of complaints about taste and discolored water from a relatively small (less than 200 properties) residential area within an urban distribution network. The area was centrally located in the distribution area and did not contain any dead ends. The predominant pipe material in this area was unlined cast iron however, unlined cast iron was common throughout the rest of the network without giving rise to similar problems.

Analysis

A network model was built of the area to analyze age of water. The model was calibrated against a tracer study using sodium chloride as the trace chemical and measuring conductivity. The model predicted a number of useful points:

- The typical velocity in the network was very low and it was suspected that this would allow sedimentation
- The pipes in the problem area were prone to flow reversal throughout the day
- The mean water age in the area was approximately 6 days which was surprisingly high given the relative proximity to the source. Furthermore, slugs of much older water (in excess of 12 days) were also detected.
- There was a significant change in consumption patterns in the summer period which could disturb sediment deposits

The fundamental problem with the area appeared to be that it was at a hydraulic balance point within the network with flow supplied from several directions, but stagnating where the flow paths meet.

Solution:

The solution was to close several valves. This rerouting of supplies increased the flow through the area thereby significantly reducing the peak water age in the problem area. However, the solution will have increased the retention time in the proximity of the shut valves.

A flushing program was set up to remove loose deposits so that they would not be disturbed by the summer demand patterns.

Key Points:

- It is generally considered good practice to avoid dead ends, however, problems can occur in looped networks if there is a hydraulic balance between two or more supply routes. Such problems can occur at locations well within the body of a network which intuitively would not be considered to be at risk.
- Age of water modeling can be useful for investigating such problems, but require accurate models. Tracer studies will greatly improve confidence in the hydraulic and age of water model calibration.

WORKSHOP CONCLUSIONS

A number of important views were established from the Workshop.

1. There was unanimous agreement that water quality can be improved by managing retention time in distribution networks (distribution mains and storage)
2. The key water quality parameters to be studied in Phase II are:
    - Chlorination and DBP formation
    - Chloramination and Nitrification
    - Discoloration
3. Detailed field measurements should be undertaken before and after intervention in order to validate the effectiveness of improved retention time on the water quality
4. Deductive reasoning and common sense can solve most problems. Complex investigations involving hydraulic and water quality modeling have their role but generally only after simple approaches have been exhausted.
5. Most small utilities do not have the resources and finances to implement complex tools (e.g. hydraulic models, CFD). Hence, the value of such tools should be demonstrated on larger utilities. This will enable small utilities to see the value of the analyses and help them justify using retention time tools.
6. A question approach (list) to identify problems and solutions and for small or resource-stretched operators would be helpful
7. There is considerable knowledge on the factors that lead to water quality changes but many gaps still remain. In particular the reactions at the pipe-water interface are not clearly understood.
8. The characteristics of problems are not well understood and are often system and sub-system specific
9. The key drivers for maintaining or improving water quality are national regulations, customer complaints (satisfaction) and operational efficiency

10. There is a need to establish baselines and fingerprints and hence change trends and violations. Irrespective of specific findings, improved system monitoring will lead to a better understanding of what is going on in distribution.
11. The long-term objective has to be Real Time Monitoring of water quality parameters within distribution
12. No universal tool or approach currently exists for solving water quality problems related to retention time
13. There are a variety of useful tools for determining retention time. However, the approach to solving water quality problems should be to:
    - Keep things simple
    - Define problems and identify issues
    - Understand how the distribution system is operated and responds to changes
    - Identify if retention time is important
    - Get the operators engaged and involved in the exercise. When the benefits can be demonstrated, they will use them.
14. Retention time may or may not be the most appropriate tool to solve a particular problem
15. There is a need for simplified tools (rules of thumb) to manage retention time for operators to follow. This will give encouragement to be involved in retention time issues and give ownership. However even using simple tools will require training.
16. Communications between different parts of the organizations were seen as important to resolving water quality problems
17. There was concern that the wide scope of the problems which could be controlled using retention time management meant that it was not feasible to produce prescriptive guidance. It was recommended that the guidance should be given primarily via case studies similar to the AwwaRF manual '*Design and operation guidelines for optimization of high-rate filtration process*' (Cleasby et al. 1989).
18. The greatest need for guidance on resolving retention time related water quality issues is for small system operators

CHAPTER 7
RETENTION TIME MANAGEMENT

GENERIC APPROACH FLOW PATH

In the preceding chapters many different analysis tools and control measures have been described that are used for investigating and resolving retention time related water quality problems. Each problem can manifest itself and be analyzed in a variety of ways using a range of tools and there can be more than one engineering or operational solution. Although these tools and measures may be used in isolation, there is significant benefit to be gained by using a structured approach to investigate and resolve the problem. Figure 7.1 shows a generalized flow path which will often be applicable.

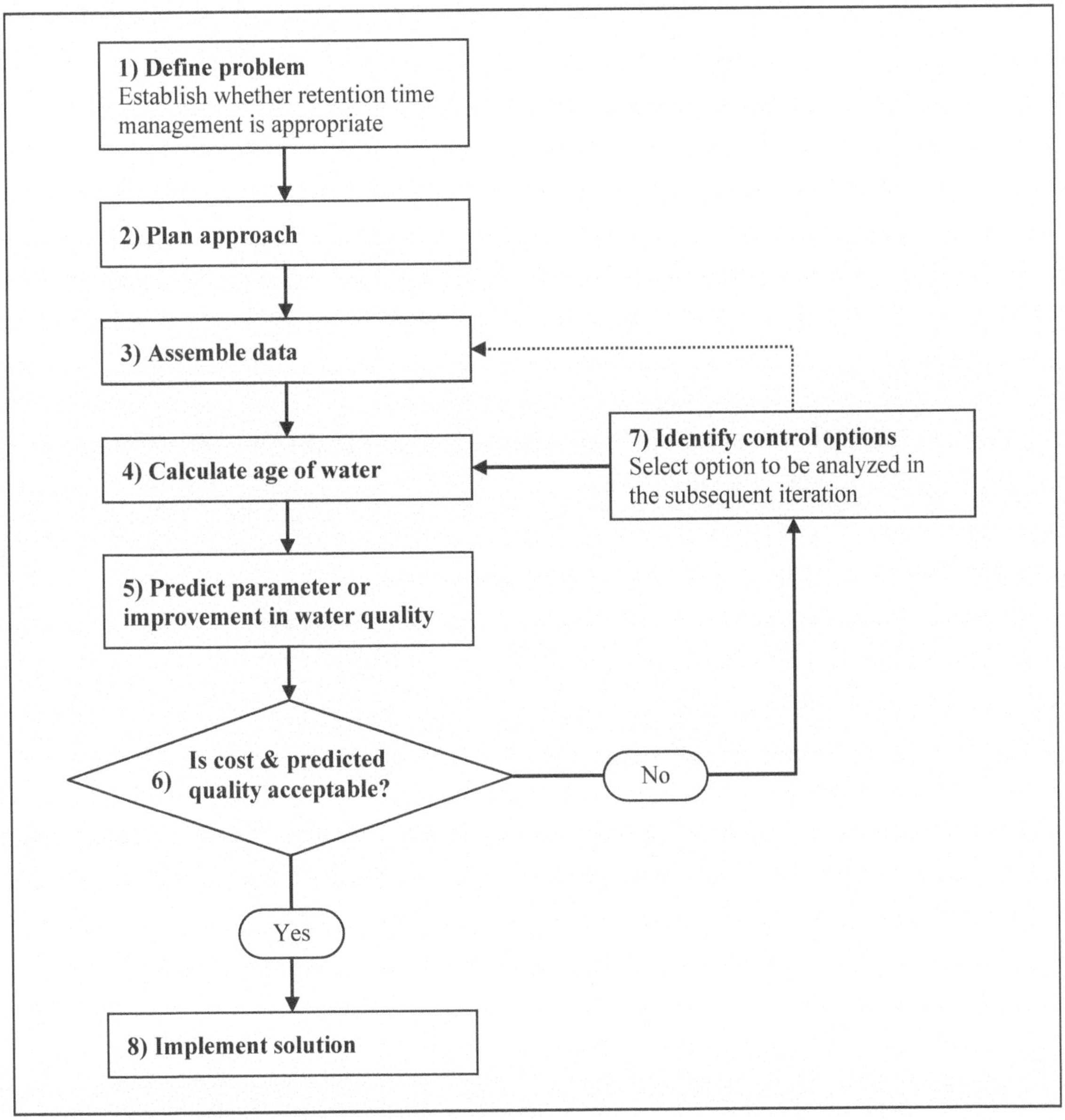

Figure 7.1 Solution cycle

Step 1 – Define Problem

At the outset of the study clearly define the problem and establish whether retention time is contributing to the problem. Then decide whether retention time management is a viable option for addressing the water quality problem taking into account any key constraints which apply. This might involve:

- Clarifying which water quality parameters are likely to be involved in the problem. This may mean going beyond the parameter which is the immediate focus of the study, for example, a microbial problem may be partly due to low chlorine residuals.
- Assessing whether these water quality parameters are influenced by retention time
- Checking that it is feasible to reduce retention time in the network. This may require detailed investigation which is not appropriate at this stage, but if it is obvious that it would not be possible to reduce retention time then alternative means of resolving the problem should be investigated.
- Giving due consideration as to whether the problem can be more effectively solved by improving water treatment or the physical condition of the distribution network pipes and facilities

The output from this step should be a clear definition of the problem, identification of the key constraints and a conclusion whether retention time management is a viable and appropriate approach to the problem.

Step 2 – Plan Approach

Plan the approach so that the data requirements and timescales can be determined. For an extensive study, it may be appropriate to develop a detailed flow chart specific to that study. This will be particularly relevant where the study involves the interaction of several different parties. A key factor for this step is the selection of the analysis methods to be used. The decisions to be taken will involve answering questions such as:

- How much time and money can be spent on the problem?
- What are the consequential costs and implications of the problem not being resolved?
- What analysis tools are appropriate for this type of problem?
- What is the availability of these tools?
- What accuracy is required?
- Are follow-on projects going to be run to address similar problems in other areas?
- What existing data are available and can they be used with suitable analysis tools?

The analysis tools selected for use and the level of accuracy required will determine the data requirements. Collecting sufficient data can take considerable time and resources and be costly. It is important therefore to plan how these data can be obtained. If the required data are not already available, their collection will often be the most expensive and time consuming part of the project. It may be appropriate to plan a multistage approach in which an initial evaluation is made using existing data followed by a more detailed analysis once sufficient data have been assembled or initial study conclusions indicate that a more detailed analysis is warranted.

Step 3 – Assemble Network Data

Utilities generally only collect the data required to meet their immediate business, operational, and regulatory needs. Additional specific data will be collected when required to meet a specific need. Due to the cost of collecting and storing data, efficient utilities are reluctant to collect and process data for which their usefulness is not established.

Water quality and related data for distribution systems are generally sparse and are not usually available in a suitable format for analyzing age of water or water quality parameters. Therefore for the majority of problems that need to be analyzed, it is unlikely that there will be sufficient data immediately available from which to derive a comprehensive solution. Consequently, either existing information will need to be manipulated and interpreted into a suitable format for the analysis or a specific data collection exercise will be required.

Information can often be interpolated from existing information using surrogate indicators; data sets that represent the performance of a particular parameter, but do not directly monitor its absolute value, and from baseline monitoring or fingerprinting using trends, changes, and violations. For some analyses, there may be adequate data already available, or assumptions can be made using the existing databases which provide an appropriate degree of accuracy for the analysis output or business risk associated with the proposed solution. This judgment will be utility or even system specific and will need to reflect the additional costs of collecting more information to the operational and business risks and implementation costs associated with the possible engineering and operational solutions.

Although there may be opportunities to collect addition information from existing facilities at low cost, no data are free and any costs associated with obtaining additional information need to be weighed up against the value of that data, including their degree of accuracy and the cost and risk implications associated with their availability.

It is important to be aware of the quality of the data that are used and how this will affect the output. No more credence should be given to the output than the quality of the input data justifies.

Step 4 – Calculate Water Age

The methods of estimating retention time are discussed in Chapter 3. These methods include manual calculations, computer modeling, physical models, and tracer studies. In the first loop of the solution cycle, the retention time corresponding to the current network setup should be estimated. However, retention time varies with time of day, season, and long-term changes. The assessment needs to be based on the conditions which are of concern. If the problem only occurs at a particular time of year then this is the condition that should be investigated.

In subsequent loops of the cycle, alternative design solutions or future scenarios are predicted. It may be that different techniques are appropriate for each loop of the cycle. Tracer studies can only be used for the current network conditions. CFD modeling would be appropriate to model the effect of changes in storage, but is generally not appropriate for calculating retention time within pipe networks.

Depending on the analysis tools available and the robustness required from the study, there are several different approaches:

1. Identify retention time at all points in the network
2. Identify retention time at key locations only

3. Identify locations with extremes in water age

Step 5 – Predict Parameter or Improvement

There are several different approaches to predicting a parameter value, including:

1. Use an integrated water quality and hydraulic model, in which case this step will be completed concurrently with the previous step.
2. Predict the water quality based on a predefined relationship with water age. This may mean using a 'batch model' or the relationship may have been derived from generic curves and tables of the characteristics of the parameter with age, conduit material, interaction with other water quality parameters, etc. The tables and curves may be representative of the utility's whole network or they may be system or subsystem specific.
3. Compare the predicted retention time with predefined threshold values which indicate the likely quality of the water. This may be based on a simple rule such as a desire to prevent water in excess of 5 days old, or may be more complex with different threshold values to indicate different levels of risk and applicable to different pipe materials and source water qualities. This approach has the merit of simplicity, but there is the difficulty of determining appropriate values. There are few existing guidelines partly because they will be system specific. Hence, threshold values for an area will not be relevant to another area unless the water type and distribution system are similar. Threshold values can be determined by a variety of methods:
    - Analysis of data collected in laboratory, pilot, or full scale studies
    - Predictions from water quality 'batch models'
    - Estimate what percentage of the network is at risk from the problem and estimate the age of water which is exceeded in the same proportion of the network
4. A percentage improvement approach: If it is not possible to predict the relationship between water quality and water age, but it is suspected that they are related, it may be applicable to assess the percentage reduction in retention time. If a significant reduction in retention time is obtained then it may be reasonable to assume that a significant, albeit unquantified, improvement in water quality would be achieved.

Step 6 – Is Cost and Predicted Quality Acceptable?

The analysis and design process needs to include assessments of whether the proposed solution will deliver the required technical performance and whether the cost of implementation is acceptable. If there is a risk that the solution will not give an acceptable water quality, an alternative solution will need to be investigated. Equally the cost of any proposed solution will need to be estimated and alternative options investigated where the cost of implementation is deemed too high.

Step 7 – Identify Control Options

Identify those control measures that could be used in the network (see Chapter 4) and select the most promising option to be analyzed in the subsequent loop of the solution cycle.

Although the solution cycle indicates that the sequence is from Step 6 to Step 7 and then Step 5, the conclusions from the analyses and solutions evolving from Step 5 may indicate that the cycle needs to revert to Step 2, to replan the approach or even restart at Step 1 by redefining the problem. The need to reappraise the whole problem is likely to result from the analysis of a complex problem where the underlying cause is masked by a more visible issue.

Step 8 – Implement Solution

When a preferred solution has been chosen it can be implemented. However, once the solution has been implemented it is prudent to continue monitoring the network to confirm that the solution is delivering the anticipated water quality improvements and hence confirm that the problem has been rectified. Ongoing monitoring will also help to build a knowledge base of the suitability of different control measures and simplify the analysis procedure if similar problems occur in the future.

CHAPTER 8
CONCLUSIONS AND RECOMMENDATIONS

PROJECT SUMMARY

The overall aim of the project is to evaluate and produce guidance on methods for managing retention time in distribution systems to control water quality. The project is in two phases:

- *Phase I*: The aim of the first Phase is to establish whether controlling retention time is a feasible option for managing water quality; identify those methods currently used for analyzing and controlling retention time; and assess their appropriateness for different circumstances.
- *Phase II*: In the second Phase, a guidance manual will be drafted on a structured approach to resolving retention time related water quality problems. It is intended that this will demonstrate how the information contained in this report can be applied appropriately. Case studies will be used to test the guidance and help demonstrate the approach through example.

This report details the studies completed during Phase I and sets out the case for using retention time management.

CONCLUSIONS

Justification for Retention Time Management

In order to demonstrate whether water quality within distribution networks can be managed by controlling retention time, the report has addressed the following three questions.

Is Water Quality Affected by Retention Time?

In Chapter 2, based on a review of literature, it was concluded that the following water quality problems are often directly related to high retention times:

- Loss of disinfectant residual
- Nitrification
- DBP formation
- Discoloration, where it is caused by stagnation in unlined iron pipes, and low velocities

Several other water quality problems are indirectly related to retention time or may be related under specific conditions

- Many tastes and odors are due to compounds in the source water however some taste and odor problems may be related to retention time, for example where it is caused by slow forming DBPs. Alternatively, chlorine may mask less acceptable tastes and hence the taste may appear to deteriorate with time as the chlorine residual decays. Similarly, any change in chlorine dose may alert a customer.

- The disinfectant residual and relatively low organic content found in most US distribution networks inhibits the growth of microorganisms. For this reason, despite numerous studies, there is little strong evidence that microbial water quality deteriorates with retention time. However, since disinfectant residual decays with retention time, the risk of microbial growth and the survival rate of any microbial contamination will, indirectly, increase with retention time.

Can Retention Time within a Distribution System be Determined?

In Chapter 3 many different techniques were presented for determining retention time. These techniques include manual calculation, computer models, tracer studies, and physical methods (summarized in Table 3.5). Furthermore, a number of methods were also presented for predicting the corresponding water quality. These methods either allow direct prediction of water quality or allow an assessment of a target retention time which will deliver acceptable quality water. The analysis tools vary in complexity. Some of the more complex water quality models need further development, simplification, and guidance on how to determine model input variables, before they will be pragmatic operational tools. However, many of the tools, such as manual calculations, tracer studies, and distribution network models, are already used by many utilities.

Can Retention Time be Controlled?

In Chapter 4 eleven different methods of controlling retention time were presented (summarized in Table 4.1). Much can be done to reduce retention time by relatively simple and inexpensive techniques including flushing at dead ends or reducing the operational top water level of reservoirs. Other techniques will involve more significant capital works such as downsizing mains. Where it is appropriate, boundary valve changes to increase velocities or shorten flow paths often provide a low cost but effective solution. Flushing is a short-term palliative that manages a problem, not a long-term solution to an underlying problem.

The methods of controlling retention time may be categorized as either system or storage related techniques. The control methods will vary in their suitability depending on the nature of the water quality problem. Water quality problems which are predominantly controlled by bulk water reactions such as THM formation can be controlled by reducing the water age either in storage facilities or within the network. However, water quality problems which are controlled by surface reactions at pipe walls are most effectively resolved by reducing the retention time in the pipes where the reaction rate is greatest.

Statement of Viability

The positive conclusions to the three questions imply that it is possible to control certain water quality parameters by managing retention time. The viability of retention time management is reinforced by the results of a survey of utilities presented in Chapter 5. This survey was completed by 23 utilities and in their response there were 68 cases where a water quality problem had been approached by using some form of retention time management as part of the solution. In the majority of cases, the utilities claimed that the solution had been successful. All the analysis tools presented in Chapter 3 were considered to be viable by at least 48 % of the utilities. At least one utility had used each of the control measures discussed in Chapter 4, and all the control measures were rated as being 'possible' or higher rating for likelihood of use by at least 50 % of the utilities.

The viability and practicality of retention time management is further reinforced by the case studies and workshop conclusions presented in Chapter 6. The case studies demonstrate that many companies have utilized retention time tools to assist in planning changes to the distribution network and resolve difficult problems. Utilities have used retention time tools to identify where in the network problems occur, solve the particular problem, and have used the experience gained to implement good operational practice.

RECOMMENDATIONS FOR A SYSTEMATIC APPROACH

Analyzing and resolving water quality problems can be complex and time consuming. Often, the process is constrained by limited data availability. Initially, the cause of the problem may be unknown and it may be unclear whether the problem is related to retention time. There is therefore a need for a comprehensive and systematic approach to solving water quality problems that also identifies whether retention time is an issue. An outline approach is proposed in Chapter 7 which will be developed into a guidance manual within the second Phase of this project.

A number of important views were recorded during the Workshop. In particular, the guidance manual will need to address the following:

1. There are a variety of useful tools for determining retention time. However, the approach to solving water quality problems should be to:
    - Keep things simple
    - Define problems and identify issues
    - Understand how the distribution system is operated and responds to changes
    - Identify if retention time is important
    - Get the operators engaged and involved in the exercise
2. Deductive reasoning and common sense can solve many problems. Hence, a question approach (list) to identify problems and solutions and for small or resource-stretched operators was recommended.
3. There is a need for simplified tools (rules of thumb) to manage retention time for operators to follow. This will encourage involvement in retention time issues and give ownership of the approach directly to the people operating the network.
4. Complex investigations involving hydraulic and water quality modeling have their role but generally after simple approaches have been exhausted

While it is desirable to reduce retention time for water quality reasons, there may be other operational management policies such as leakage reduction which result in increased retention time. Furthermore, requirements for strategic storage for firefighting and other emergencies conflict with strategies for high turnover and low retention times. Controlling retention time will affect other operational or service level requirements of the network. These should be considered prior to the implementation of a proposed change to determine whether the performance characteristics will remain within acceptable limits. In particular, many of the methods will reduce the hydraulic or storage capacity within the network, an issue that needs to be balanced against fire flow and supply pressure requirements. The guidance manual needs to address these constraints.

In order to incorporate case studies to test and illustrate the guidance, the publication of the manual is currently scheduled for 2005.

APPENDIX A
UTILITY SURVEY

The following survey was completed by the utilities listed below:

1. Amsterdam Water Supply, Netherlands
2. Brisbane Water, Australia
3. Champlain Water District
4. City of Philadelphia, Water Department
5. Denver Water
6. Duinwaterbedryf Zuid, Netherlands
7. East Bay Municipal Utility District
8. Elizabethtown Water Company
9. Greater Cincinnati Water Works
10. Greenville Water System
11. Louisville Water Company
12. Melbourne Water, Australia
13. Power & Water Authority, Darwin, NT, Australia
14. South Australia Water Corporation, Australia
15. South East Water Limited, Australia
16. Sydney Water, Australia
17. South Central Connecticut Regional Water Authority
18. Thames Water Utilities, UK
19. United Water International, Australia
20. Vitens Friesland, Netherlands
21. Vitens Overijssel BV, Netherlands
22. WLN (Waterbedryl Groningen & WMD)
23. Yorkshire Water, UK

EVALUATING RETENTION TIME TO MANAGE DISTRIBUTION SYSTEM WATER QUALITY: SURVEY OF UTILITY PRACTICES

INTRODUCTION

When water leaves a treatment works and travels through a distribution network, its quality, with respect to many parameters, will degrade. The mechanisms by which this occurs are very complex and will be affected by:

- The full chemical and biological makeup of the water, not just the concentrations of the individual parameters of concern
- The mixing of water from different sources
- Hydraulic characteristics
- Dispersion characteristics
- Temperature
- Time
- The material and condition of any surface in contact with the water, most notably this will be the internal pipe wall
- Physical and operational characteristics which result in contamination

Table A.1 summarizes some of these water quality issues. Models have been produced to analyze many water quality parameters within distribution. These inevitably make assumptions to simplify the mechanisms. While some of these models have given good results, they are often still too complex to be applied as pragmatic operational tools and may require data which is not readily available to operators. Nevertheless, the quality of the delivered water will be largely influenced by:

- The quality of treated water
- The condition of distribution assets
- Retention time

To date utilities have predominantly focused on the quality of treated water and the condition of distribution assets when managing water quality. Retention time is a consequence of physical characteristics of the system and operational activities within it. Physical characteristics may change over the life of the asset or be modified by rehabilitation. Operational activities may be structured (e.g. pump scheduling and planned maintenance) or uncontrolled (e.g. demand variations). This project looks at the potential for managing the retention time to control water quality.

Table A.1
Specific water quality parameters affected by distribution systems

Problem	Monitoring Parameter	Potential causes
Loss of disinfectant residual	Disinfectant residual (chloramines, nitrite, and HPCs)	• Long retention time • High levels of TOC in finished water • Excessive amounts of unlined cast-iron pipe
Microbial	Coliforms (total and fecal) HPCs	• Reduction in disinfectant residual due to long retention time • Poor source water treatment • Intrusion • Unlined cast-iron pipe
Discolored water	Iron, manganese, turbidity, color and consumer complaints	• Unlined cast-iron pipe • Deposition and accumulation of sediment at low flow velocities which can then be resuspended if velocities increase • Long retention time in unlined cast-iron pipe • Poor chemistry (e.g. low alkalinity) • Carry over of iron or other compounds from treatment
Disinfection by-product formation	THMs, HAAs	• Long retention time • Poor pH control • Poor source water treatment.
Corrosion of metals – lead, iron, and copper	Pb, Cu, Fe	• Poor chemical treatment (e.g. pH, carbonate, and orthophosphate) • Long standing times in pipe • Long retention time may reduce pH which could increase corrosion
Taste and odor	Phenolic compounds Cl_2 residual Consumer complaints	• Long standing times with high chlorine, high pH, and NOM • Long standing times with low chlorine and microbial activity
Nitrification	Nitrite, HPCs, ammonia	• Excessive retention time causing release of ammonia • Dosing ammonia in excess • Poor source water • Treatment high TOC • Nitrification is inhibited at low pH

Many utilities are likely to have implemented some form of retention time management techniques without specifically classifying it as 'retention time management'. Much can be done to reduce retention time by relatively simple and inexpensive techniques such as flushing at dead ends or reducing the operational top water level of reservoirs. Other techniques will involve more significant capital works. Table A.2 outlines some of the methods that can be used to control retention time.

Table A.2
Methods for controlling retention time.

Method	Details
Altering the valving of networks	By changing valve arrangements and hydraulic boundaries, travel times can be reduced and water rerouted to increase velocities in low flowing pipes to maximize velocities
Installing time varying valves	Time and other control valves can be used to control flows
Manual flushing	Periodic flushing to remove sediment and reduce water age in dead ends and low velocity sections of pipe
Automated flushing	Automatic programmed flushing to remove sediment and reduce water age in dead ends and low velocity sections of pipe
Abandoning mains	Remove surplus capacity from system
Downsizing mains	Reduce system capacity to increase velocities and reduce retention times
Adjusting pump schedules	Optimize pumping regime to match supply and demand, minimize energy requirements and retention times (linked to optimizing system storage)
Increasing turnover of reservoirs	Eliminate or reduce strategic storage and manage diurnal storage to minimize retention time. This is likely to be linked to resource capacity and operational optimization of both resources and pumping facilities
Reducing the operational top water level (TWL) of reservoirs	Eliminate or reduce strategic storage and manage diurnal storage to minimize retention time. The TWL may be varied with seasons
Taking reservoirs out of service or removing reservoir cells	Remove surplus capacity from system
Altering reservoir configuration to avoid dead zones	Replace common inlet outlet with separate pipes or install baffles to eliminate dead water zones and improve throughput thereby reduce retention time

If a utility has a water quality problem, then they could try some of the simpler techniques for reducing retention time and see if an improvement is obtained. However, it is often not straightforward as to how to apply the techniques in the optimal fashion and if any significant expense is involved some form of assessment of the potential benefit is likely to be required. Hence, there is a need for tools to assess retention time and how it will be affected by any proposed changes. Table A.3 outlines some such tools.

Table A.3
Analysis tools.

Method	Details
Manual calculations	Simple systems can be analyzed using manual and spreadsheet calculations. Pipe flow equations combined with storage or turnover assessments can be used to derive retention time estimates for snapshot conditions and simple system extended periods. Age is then compared with water quality records and customer complaints to identify parts of systems prone or liable to quality failure (chemical and biological).
Rules of thumb or experience	Generic rules are used together with hydraulic assessment and an understanding of system operations (e.g. for a specific Cl_2 dose, expect stagnant water complaints where age is over 100 hours). Operators' practical and specific experience can be used to identify parts of their system prone to water quality problems (e.g. under certain demand or supply conditions, problems may occur at known locations).
Tracer studies	By injecting a chemical into a network and then monitoring for it at a downstream point, the retention time can be directly measured. This technique is relatively straightforward, especially when the trace chemical is already present in the system and can be easily adjusted e.g. fluoride or chlorine.
Review of water quality data	Review of water quality data can indicate areas with water quality problems. These may then correlate with areas of high water age.
GIS (geographical information systems)	GIS can be used to select those properties which are at the extremities of the network or near dead ends and so vulnerable to a high water age. It is also possible to incorporate more sophisticated techniques by assigning characteristic water quality properties (including retention time) to each element in the GIS. Queries can then be set up which follow rules to estimate the water quality within the network. GIS can also be used to present database information spatially to look for patterns and causes of problems.
Hydraulic network model	Computer models of the hydraulic behavior of distribution networks such as EPANET, H2Onet, Stoner, WaterCad, Aquis, and Infoworks. These software can be used to predict the water age and highlight any areas in which it would exceed a given value.
Water quality network model	Many of the above network models can also be used to model water quality for parameters whose concentration is a direct function of water age. The most commonly modeled parameter is chlorine. However, this does require a reliable estimate of reaction coefficients which may not be readily available.
Water quality statistical model	Some utilities have developed models of water quality, which are purely derived from statistical analysis of the vast quantities of water quality data that are collected for regulatory purposes
Reservoir model	Assessments of turnover including strategic storage, diurnal, and seasonal usage give indications of retention time and hence the preponderance of water quality problems. However 3-D/CFD modeling may still be needed to understand internal circulation, dead zones, etc.
Physical model	An expensive tool, not normally used for system operational management. However it may be useful for validating new design (e.g. the location of baffles to prevent dead zones in reservoirs).

(continued)

Table A.3 (continued)

Method	Details
CFD models (computational fluid dynamics)	Computational fluid dynamics is increasingly being used to analyze the behavior of water within tanks and reservoirs. It can highlight dead zones in which there is limited turnover of water. Methods for rectifying problems such as installing baffles can also be modeled. CFD models have also been used to simulate flow in pipes including at dead ends under flushing conditions.
Taste tests	Taste can be indicative of high retention times. Carefully designed and executed taste testing exercises can provide useful insight into particular problems.
Customer perception surveys	Carefully designed and executed surveys can provide useful information and insight into particular problems. However, they need to be used in conjunction with other tools to provide information on retention time issues. They have been widely used in the UK in conjunction with structured sampling to identify and justify network improvement schemes.
Neural network models	Neural networks are potentially of benefit where it is reasonable to assume that there is a relationship between parameters, but where the mechanism by which they are related is not understood. They are effectively used to derive empirical relationships and could be used in a similar method to statistical water quality models (see above). To date they have not been widely used.
Optimization software	Optimization tools are widely available within spreadsheets (e.g. the 'Solver' add-in within Microsoft Excel) and can be used to optimize many problems which can be defined in terms of mathematical functions. Optimization tools are also increasingly being incorporated into other analysis software such as network models.

While it is desirable to reduce retention time for water quality reasons, there may be other policies which can result in increased retention time:

- Leakage reduction
- Zoning for metering purposes
- Maximizing hydraulic capacity

The benefits of these policies can be readily assessed and quantified. Unless tools are available for assessing their impact on retention time, it is not possible to reliably compare the benefits with any adverse effects on water quality and so this is likely to be overlooked.

SURVEY OF UTILITY PRACTICES

The starting point for this AwwaRF study is to identify to what extent retention time management is currently used to manage water quality. To do this we are circulating this questionnaire to personnel in the water industry who are likely to be involved in such work. We would be very grateful if you could take the time to supply us with information as it will help focus the study onto the current needs of the water industry. Even if your utility does none of the practices outlined in the questionnaire it would be of benefit to know this.

1 Company details

A	Name of person/s completing survey			
B	Name of company			
C	Size of company e.g. Number of properties served			
D	Please tick the issues that are or have been relevant to your company then complete the relevant question sheets			*If ticked please complete:*
		Loss of disinfectant residual	☐	*Question sheet 2*
		High microbial counts	☐	*Question sheet 3*
		Animal infestation	☐	*Question sheet 4*
		Discolored water	☐	*Question sheet 5*
		Disinfection by-products (DBPs)	☐	*Question sheet 6*
		Leaching of PACs from bitumen lined pipes	☐	*Question sheet 7*
		Corrosion of metals	☐	*Question sheet 8*
		Taste and odor	☐	*Question sheet 9*
		Nitrification	☐	*Question sheet 10*
		Any other parameters	☐	*Question sheet 11*
E	Please now complete the two remaining question sheets: 12 - Feasibility of retention time management techniques. 13 - Feasibility of retention time management analysis tools.			
C	Thank you for participating in this survey. The results of this study will be incorporated into an AwwaRF report. We will be looking for case studies of various forms for inclusion in this report. Would you be willing for us to contact you about this?			YES/NO

2 **Loss of disinfectant residual** e.g. free chlorine or chloramine

A	Which parameters are a problem? (Please state e.g. free chlorine or chloramines)				
B	How was the problem identified?	regulatory sampling	☐	specific investigation	☐
		customer complaints	☐		
		Other (please state)	----------		
C	What was the primary cause of the problem? (Please state e.g. unlined iron pipes or high TOC)				
D	How did you rectify the problem? (Please state e.g. increased chlorine dose or not rectified)				
E	Have you undertaken any work to address the problem by controlling retention time?	YES/NO		***If NO go to F, if YES go to G,***	
F	If NO to question E, why? ***Then return to question sheet 1***	Not considered at start	☐	Not confident it would work	☐
		Cost	☐		
		Other (please state)	----------		
G	If YES to question E, what method was used to alter the retention time? ***See table 2 for more details***	Altering valving of network	☐	Installing time varying valves	☐
		Manual flushing	☐	Automated flushing	☐
		Abandoning mains	☐	Downsizing mains	☐
		Adjusting pump schedules	☐	Reducing TWL of reservoirs	☐
		Removing reservoir cells	☐	Altering reservoir configuration	☐
		Other (please state)	----------		
H	If YES to question E, what method was used to analyze and plan the work. ***See table 3 for more details***	None	☐	Manual calculations	☐
		Rules of thumb / experience	☐	Review of water quality data	☐
		GIS	☐	Hydraulic network model	☐
		Water quality network model	☐	Water quality statistical model	☐
		Reservoir model	☐	Physical model	☐
		CFD model	☐	Taste Tests	☐
		Customer perception survey	☐	Optimization software	☐
		Other (please state)	----------		
I	If YES to question E, has the work improved water quality?	No - perception	☐	No – demonstrated by data	☐
		Yes - perception	☐	Yes – demonstrated by data	☐

3	**High microbial counts** e.g. coliform or total plate counts				
A	Which parameters are a problem? (Please state e.g. coliform or total plate counts)	☐			
B	How was the problem identified?	regulatory sampling	☐	specific investigation	☐
		customer complaints	☐		
		Other (please state) ----------			
C	What was the primary cause of the problem? (insufficient disinfectant residual)	☐			
D	How did you rectify the problem? (Please state e.g. increased chlorine dose or not rectified)	☐			
E	Have you undertaken any work to address the problem by controlling retention time?	YES/NO		***If NO go to F, if YES go to G,***	
F	If NO to question E, why? ***Then return to question sheet 1***	Not considered at start	☐	Not confident it would work	☐
		Cost	☐		
		Other (please state) ----------			
G	If YES to question E, what method was used to alter the retention time? ***See table 2 for more details***	Altering valving of network	☐	Installing time varying valves	☐
		Manual flushing	☐	Automated flushing	☐
		Abandoning mains	☐	Downsizing mains	☐
		Adjusting pump schedules	☐	Reducing TWL of reservoirs	☐
		Removing reservoir cells	☐	Altering reservoir configuration	☐
		Other (please state) ----------			
H	If YES to question E, what method was used to analyze and plan the work. ***See table 3 for more details***	None	☐	Manual calculations	☐
		Rules of thumb / experience	☐	Review of water quality data	☐
		GIS	☐	Hydraulic network model	☐
		Water quality network model	☐	Water quality statistical model	☐
		Reservoir model	☐	Physical model	☐
		CFD model	☐	Taste Tests	☐
		Customer perception survey	☐	Optimization software	☐
		Other (please state) ----------			
I	If YES to question E, has the work improved water quality?	No - perception	☐	No – demonstrated by data	☐
		Yes - perception	☐	Yes – demonstrated by data	☐

4 **Animal infestation** e.g. Asselus or snails

A	Which parameters are a problem? (Please state e.g. Asselus or snails)				
B	How was the problem identified?	regulatory sampling	☐	specific investigation	☐
		customer complaints	☐		
		Other (please state) ----------			
C	What was the primary cause of the problem? (Please state e.g. high organic content in water)				
D	How did you rectify the problem? (Please state e.g. flushing or not rectified)				
E	Have you undertaken any work to address the problem by controlling retention time?	YES/NO		*If NO go to F, if YES go to G,*	
F	If NO to question E, why? ***Then return to question sheet 1***	Not considered at start	☐	Not confident it would work	☐
		Cost	☐		
		Other (please state) ----------			
G	If YES to question E, what method was used to alter the retention time? ***See table 2 for more details***	Altering valving of network	☐	Installing time varying valves	☐
		Manual flushing	☐	Automated flushing	☐
		Abandoning mains	☐	Downsizing mains	☐
		Adjusting pump schedules	☐	Reducing TWL of reservoirs	☐
		Removing reservoir cells	☐	Altering reservoir configuration	☐
		Other (please state) ----------			
H	If YES to question E, what method was used to analyze and plan the work. ***See table 3 for more details***	None	☐	Manual calculations	☐
		Rules of thumb / experience	☐	Review of water quality data	☐
		GIS	☐	Hydraulic network model	☐
		Water quality network model	☐	Water quality statistical model	☐
		Reservoir model	☐	Physical model	☐
		CFD model	☐	Taste Tests	☐
		Customer perception survey	☐	Optimization software	☐
		Other (please state) ----------			
I	If YES to question E, has the work improved water quality?	No - perception	☐	No – demonstrated by data	☐
		Yes - perception	☐	Yes – demonstrated by data	☐

5	**Discolored water** Including red water / high iron concentrations				
A	Which parameters are a problem? (Please state e.g. iron sediment, manganese or chalk)	*Alternatively describe the appearance*			
B	How was the problem identified?	regulatory sampling	☐	specific investigation	☐
		customer complaints	☐		
		Other (please state) ------------------------------			
C	What was the primary cause of the problem? (Please state e.g. treatment process or iron pipes)				
D	How did you rectify the problem? (Please state e.g. flushing or not rectified)				
E	Have you undertaken any work to address the problem by controlling retention time?	YES/NO		***If NO go to F, if YES go to G,***	
F	If NO to question E, why? ***Then return to question sheet 1***	Not considered at start	☐	Not confident it would work	☐
		Cost	☐		
		Other (please state) ------------------------------			
G	If YES to question E, what method was used to alter the retention time? ***See table 2 for more details***	Altering valving of network	☐	Installing time varying valves	☐
		Manual flushing	☐	Automated flushing	☐
		Abandoning mains	☐	Downsizing mains	☐
		Adjusting pump schedules	☐	Reducing TWL of reservoirs	☐
		Removing reservoir cells	☐	Altering reservoir configuration	☐
		Other (please state) ------------------------------			
H	If YES to question E, what method was used to analyze and plan the work. ***See table 3 for more details***	None	☐	Manual calculations	☐
		Rules of thumb / experience	☐	Review of water quality data	☐
		GIS	☐	Hydraulic network model	☐
		Water quality network model	☐	Water quality statistical model	☐
		Reservoir model	☐	Physical model	☐
		CFD model	☐	Taste Tests	☐
		Customer perception survey	☐	Optimization software	☐
		Other (please state) ------------------------------			
I	If YES to question E, has the work improved water quality?	No - perception	☐	No – demonstrated by data	☐
		Yes - perception	☐	Yes – demonstrated by data	☐

6 **Disinfection by-products (DBPs)** e.g. Trihalomethanes (THMs)

A	Which parameters are a problem? (Please state e.g. Trihalomethanes)				
B	How was the problem identified?	regulatory sampling	☐	specific investigation	☐
		customer complaints	☐		
		Other (please state) ----------			
C	What was the primary cause of the problem? (Please state e.g. high organic content in water)				
D	How did you rectify the problem? (Please state e.g. installation of GAC treatment)				
E	Have you undertaken any work to address the problem by controlling retention time?	YES/NO		***If NO go to F, if YES go to G,***	
F	If NO to question E why? ***Then return to question sheet 1***	Not considered at start	☐	Not confident it would work	☐
		Cost	☐		
		Other (please state) ----------			
G	If YES to question E, what method was used to alter the retention time? ***See table 2 for more details***	Altering valving of network	☐	Installing time varying valves	☐
		Manual flushing	☐	Automated flushing	☐
		Abandoning mains	☐	Downsizing mains	☐
		Adjusting pump schedules	☐	Reducing TWL of reservoirs	☐
		Removing reservoir cells	☐	Altering reservoir configuration	☐
		Other (please state) ----------			
H	If YES to question E, what method was used to analyze and plan the work. ***See table 3 for more details***	None	☐	Manual calculations	☐
		Rules of thumb / experience	☐	Review of water quality data	☐
		GIS	☐	Hydraulic network model	☐
		Water quality network model	☐	Water quality statistical model	☐
		Reservoir model	☐	Physical model	☐
		CFD model	☐	Taste Tests	☐
		Customer perception survey	☐	Optimization software	☐
		Other (please state) ----------			
I	If YES to question E, has the work improved water quality?	No - perception	☐	No – demonstrated by data	☐
		Yes - perception	☐	Yes – demonstrated by data	☐

7 Leaching of PAH from bitumen lined pipes

A	Which parameters are a problem? (Please state)				
B	How was the problem identified?	regulatory sampling	☐	specific investigation	☐
		customer complaints	☐		
		Other (please state)	----------		
C	What was the primary cause of the problem? (Please state e.g. pipe linings)				
D	How did you rectify the problem? (Please state e.g. pipe replacement)				
E	Have you undertaken any work to address the problem by controlling retention time?	YES/NO		***If NO go to F if YES go to G***	
F	If NO to question E why? ***Then return to question sheet 1***	Not considered at start	☐	Not confident it would work	☐
		Cost	☐		
		Other (please state)	----------		
G	If YES to question E, what method was used to alter the retention time? ***See table 2 for more details***	Altering valving of network	☐	Installing time varying valves	☐
		Manual flushing	☐	Automated flushing	☐
		Abandoning mains	☐	Downsizing mains	☐
		Adjusting pump schedules	☐	Reducing TWL of reservoirs	☐
		Removing reservoir cells	☐	Altering reservoir configuration	☐
		Other (please state)	----------		
H	If YES to question E, what method was used to analyze and plan the work. ***See table 3 for more details***	None	☐	Manual calculations	☐
		Rules of thumb / experience	☐	Review of water quality data	☐
		GIS	☐	Hydraulic network model	☐
		Water quality network model	☐	Water quality statistical model	☐
		Reservoir model	☐	Physical model	☐
		CFD model	☐	Taste Tests	☐
		Customer perception survey	☐	Optimization software	☐
		Other (please state)	----------		
I	If YES to question E, has the work improved water quality?	No - perception	☐	No – demonstrated by data	☐
		Yes - perception	☐	Yes – demonstrated by data	☐

8	**Corrosion of metals** e.g. lead, iron and copper				
A	Which parameters are a problem? (Please state e.g. lead, iron copper)				
B	How was the problem identified?	regulatory sampling	☐	specific investigation	☐
		customer complaints	☐		
		Other (please state) ----------			
C	What was the primary cause of the problem? (Please state e.g. unlined pipes)				
D	How did you rectify the problem? (Please state e.g. flushing or not rectified)				
E	Have you undertaken any work to address the problem by controlling retention time?	YES/NO		***If NO go to F if YES go to G***	
F	If NO to question E why? ***Then return to question sheet 1***	Not considered at start	☐	Not confident it would work	☐
		Cost	☐		
		Other (please state) ----------			
G	If YES to question E, what method was used to alter the retention time? ***See table 2 for more details***	Altering valving of network	☐	Installing time varying valves	☐
		Manual flushing	☐	Automated flushing	☐
		Abandoning mains	☐	Downsizing mains	☐
		Adjusting pump schedules	☐	Reducing TWL of reservoirs	☐
		Removing reservoir cells	☐	Altering reservoir configuration	☐
		Other (please state) ----------			
H	If YES to question E, what method was used to analyze and plan the work. ***See table 3 for more details***	None	☐	Manual calculations	☐
		Rules of thumb / experience	☐	Review of water quality data	☐
		GIS	☐	Hydraulic network model	☐
		Water quality network model	☐	Water quality statistical model	☐
		Reservoir model	☐	Physical model	☐
		CFD model	☐	Taste Tests	☐
		Customer perception survey	☐	Optimization software	☐
		Other (please state) ----------			
I	If YES to question E, has the work improved water quality?	No - perception	☐	No – demonstrated by data	☐
		Yes - perception	☐	Yes – demonstrated by data	☐

9 Taste and odor

A	Which parameters are a problem? (Please state e.g. earth-like or specific compound if identified)				
B	How was the problem identified?	regulatory sampling	☐	specific investigation	☐
		customer complaints	☐		
		Other (please state)	----------		
C	What was the primary cause of the problem? (Please state e.g. high chlorine residual)				
D	How did you rectify the problem? (Please state e.g. reduced chlorine dose)				
E	Have you undertaken any work to address the problem by controlling retention time?	YES/NO		***If NO go to F if YES go to G***	
F	If NO to question E why? ***Then return to question sheet 1***	Not considered at start	☐	Not confident it would work	☐
		Cost	☐		
		Other (please state)	----------		
G	If YES to question E, what method was used to alter the retention time? ***See table 2 for more details***	Altering valving of network	☐	Installing time varying valves	☐
		Manual flushing	☐	Automated flushing	☐
		Abandoning mains	☐	Downsizing mains	☐
		Adjusting pump schedules	☐	Reducing TWL of reservoirs	☐
		Removing reservoir cells	☐	Altering reservoir configuration	☐
		Other (please state)	----------		
H	If YES to question E, what method was used to analyze and plan the work. ***See table 3 for more details***	None	☐	Manual calculations	☐
		Rules of thumb / experience	☐	Review of water quality data	☐
		GIS	☐	Hydraulic network model	☐
		Water quality network model	☐	Water quality statistical model	☐
		Reservoir model	☐	Physical model	☐
		CFD model	☐	Taste Tests	☐
		Customer perception survey	☐	Optimization software	☐
		Other (please state)	----------		
I	If YES to question E, has the work improved water quality?	No - perception	☐	No – demonstrated by data	☐
		Yes - perception	☐	Yes – demonstrated by data	☐

10 Nitrification

A	Which parameters are a problem? (Please state)				
B	How was the problem identified?	regulatory sampling	☐	specific investigation	☐
		customer complaints	☐		
		Other (please state) ----------			
C	What was the primary cause of the problem? (Please state e.g. high water age)				
D	How did you rectify the problem? (Please state e.g. flushing or not rectified)				
E	Have you undertaken any work to address the problem by controlling retention time?	YES/NO		***If NO go to F if YES go to G***	
F	If NO to question E why? ***Then return to question sheet 1***	Not considered at start	☐	Not confident it would work	☐
		Cost	☐		
		Other (please state) ----------			
G	If YES to question E, what method was used to alter the retention time? ***See table 2 for more details***	Altering valving of network	☐	Installing time varying valves	☐
		Manual flushing	☐	Automated flushing	☐
		Abandoning mains	☐	Downsizing mains	☐
		Adjusting pump schedules	☐	Reducing TWL of reservoirs	☐
		Removing reservoir cells	☐	Altering reservoir configuration	☐
		Other (please state) ----------			
H	If YES to question E, what method was used to analyze and plan the work. ***See table 3 for more details***	None	☐	Manual calculations	☐
		Rules of thumb / experience	☐	Review of water quality data	☐
		GIS	☐	Hydraulic network model	☐
		Water quality network model	☐	Water quality statistical model	☐
		Reservoir model	☐	Physical model	☐
		CFD model	☐	Taste Tests	☐
		Customer perception survey	☐	Optimization software	☐
		Other (please state) ----------			
I	If YES to question E, has the work improved water quality?	No - perception	☐	No – demonstrated by data	☐
		Yes - perception	☐	Yes – demonstrated by data	☐

11 Any other parameters

A	Which parameters are a problem? (Please state)				
B	How was the problem identified?	regulatory sampling	☐	specific investigation	☐
		customer complaints	☐		
		Other (please state) ------------------------------			
C	What was the primary cause of the problem? (Please state e.g. high organic content in water)				
D	How did you rectify the problem? (Please state e.g. flushing or not rectified)				
E	Have you undertaken any work to address the problem by controlling retention time?	YES/NO		***If NO go to F, if YES go to G,***	
F	If NO to question C, why? ***Then return to question sheet 1***	Not considered at start	☐	Not confident it would work	☐
		Cost	☐		
		Other (please state) ------------------------------			
G	If YES to question E, what method was used to alter the retention time? ***See table 2 for more details***	Altering valving of network	☐	Installing time varying valves	☐
		Manual flushing	☐	Automated flushing	☐
		Abandoning mains	☐	Downsizing mains	☐
		Adjusting pump schedules	☐	Reducing TWL of reservoirs	☐
		Removing reservoir cells	☐	Altering reservoir configuration	☐
		Other (please state) ------------------------------			
H	If YES to question E, what method was used to analyze and plan the work. ***See table 3 for more details***	None	☐	Manual calculations	☐
		Rules of thumb / experience	☐	Review of water quality data	☐
		GIS	☐	Hydraulic network model	☐
		Water quality network model	☐	Water quality statistical model	☐
		Reservoir model	☐	Physical model	☐
		CFD model	☐	Taste Tests	☐
		Customer perception survey	☐	Optimization software	☐
		Other (please state) ------------------------------			
I	If YES to question E, has the work improved water quality?	No - perception	☐	No – demonstrated by data	☐
		Yes - perception	☐	Yes – demonstrated by data	☐

12	**Feasibility of retention time management techniques**		
	The purpose of this section is to identify those management techniques which the water industry considers to be pragmatic and effective.		
	Technique. ***See Table 2 for more details about these techniques***	On a scale of 1 to 5 how likely is it that this technique could be used in your company within the next 5 years? 1 = No chance 2 = unlikely 3 = possible 4 = likely 5 = Already in use	Please give any comments about the viability of the technique.
A	Altering the valving of networks		
B	Installing time varying valves		
C	Manual flushing		
D	Automated flushing		
E	Abandoning mains		
F	Downsizing mains		
G	Adjusting pump schedules		
H	Reducing the operational top water level (TWL) of reservoirs		
I	Removing reservoir cells		
J	Altering reservoir configuration to avoid dead zones		

H Please list below any other techniques for controlling retention time of which you are aware.

13 Feasibility of retention time management analysis tools

The purpose of this section is to identify those management analysis tools which the water industry considers to be pragmatic and effective.

	Analysis tool. ***See Table 2 for more details about these techniques***	Is this a viable tool for your company?	Has your company used this tool?	Please give any relevant details e.g. software used.
A	Manual calculations	YES/NO	YES/NO	
B	Rules of thumb / experience	YES/NO	YES/NO	
C	Tracer studies	YES/NO	YES/NO	What have you used as the trace chemical?
D	Review of water quality data	YES/NO	YES/NO	
E	GIS	YES/NO	YES/NO	
F	Hydraulic network model	YES/NO	YES/NO	
G	Water quality network model	YES/NO	YES/NO	
H	Water quality statistical model	YES/NO	YES/NO	
I	Reservoir model	YES/NO	YES/NO	
J	Physical model	YES/NO	YES/NO	
K	CFD models	YES/NO	YES/NO	
L	Taste tests	YES/NO	YES/NO	
M	Customer perception surveys	YES/NO	YES/NO	
N	Neural network models	YES/NO	YES/NO	
O	Optimization software	YES/NO	YES/NO	
P	Please list any other relevant analysis techniques of which you are aware.			

REFERENCES

Ackers, J., M. Brandt, and J. Powell. 2001. *Hydraulic Characterisation of Deposits and Review of Sediment Modelling*. London, England: UK Water Industry Research Limited (UKWIR).

Amy, G.L., P.A. Chadik, and Z.K. Chowdhary. 1987. Developing Models for Predicting Trihalomethane Formation Potential and Kinetics. *Journal of the American Water Works Association*, 79(7):89-97.

Anon. 1996. *Australian Drinking Water Quality Guidelines* (updated September 2001). Published by the National Health and Medical Research Council and the Agriculture and Resource Management Council of Austrailia and New Zealand.

Anon. 1998. Council Directive 98/83/EC on the Quality of Water Intended for Human Consumption. Official Journal of the European Communities L330/32.

APHA (American Public Health Association). 1994. *Standard Methods for the Examination of Water and Wastewater.* Washington, D.C. APHA.

AwwaRF and DVGW (Deutscher Verein des Gas and Wasserfachen). 1985. *Internal Corrosion of Water Distribution Systems*. Denver, Colo.: AwwaRF and AWWA.

Benrahnem, M. 1996. Validation par un Tracage des Temps de Sejour de l'eau Dans Reseau de Distrituion d'eau Potable. *Memoire de Maitrise.* Ecole Polytechique de Montreal. 209p. (cited in Desjardins et al 2000).

Block, J-C., K. Haudidier, J.L. Paquin, J. Miazga, Y. Levi, and J.D. Bryers. 1993. Biofilm Accumulation in Drinking Water Distribution Systems. *Biofouling* 6(4):333-343.

Boulos, P.F., T. Altman, P.A. Jarrige, and F. Collevati. 1994. An Event Driven Method for Modelling Contaminant Propagation in Water Networks. *Applied Mathematical Modelling*, 18:84-92.

Boulos, P.F., T. Altman, P.A. Jarrige, and F. Collevati. 1995. Discrete Simulation Approach for Network-Water-Quality Models. *Journal of Water Resources Planning and Management,* 121(1):49-60.

Brammer, L.F., and N.S. Tansley. 1993. Chlorine Residual Modelling in Distribution: The Improvement of Taste and the Maintenance of Effective Disinfection. Presented at Computer Applications for Water Supply and Distribution Conference, DeMontfort University, Leicester, UK, September 7-9, 1993.

Brandt, M. 2000. Taste Action Plan, Customer Perception of Drinking Water Taste. Presented at CIWEM Metropolitan branch meeting, London. January 26, 2000.

Brandt, M., J. Creasey, and R. Perkins. 1999. Making Decisions that Matter with Water Quality Models. In *Proc. of the International Conference on Computing and Control for the Water Industry, 13-14 september 1999.* Hertfordshire, UK: Research Studies Press.

Brendan, B.L., and G. Kuczera. 1997. Network Linear Programming as Pipe Network Hydraulic Analysis Tool. *Journal of Hydraulic Engineering,* 123 (6):549-559.

Burgess, P.L., D.C. Vitanage, A. Holmes, and P.J. Bliss. 1996. Determination of the Importance of Wall Effects in Modelling Chlorine Residuals within Distribution Systems. In *Proc. of the Australian Water and Wastewater Assocaition 16th Federal Convention.* Sydney, Australia: Australian Water and Wastewater Assocaition.

Burlingame, G.A., and M. Johnson. 2002. Get More for Your NTUs. *Opflow*, 18(12):8-9. Denver, Colo.: AWWA.

Camper, A.K., J.T. Hayes, W.L. Jones, and N. Zelver. 1993. Persistence of Coliforms in Mixed Population Biofilms. In *Proc. of the AWWA Annual Water Quality Technology Conference, Miami, November 7-11, 1993.* Denver, Colo.: AWWA.

Camper, A.K. 1996. *Factors Limiting Microbial Growth in Distribution Systems: Laboratory and Pilot-Scale Experiments.* Denver, Colo.: AwwaRF and AWWA.

Cardew, P., W. Grayman, J. Marshall, and P. Marshall. 2000. *Aspects of Modelling Chlorine Decay in Water Distribution Systems.* London, England: UK Water Industry Research Limited (UKWIR).

Cesario, A.L., J.R. Kroon, and W Grayman. 1996. New Perspectives on Calibration of Treated Water Distribution System Models. In *Proc. of 1996 AWWA Annual Conference.* Denver, Colo.: AWWA.

CFX. 2001. *Tanksim.* CFD Tank Modelling Software. Sheffield, UK: CFX.

Chambers, K., R. Crane, J. Creasey, J. Marshall, N. Mole, R. Pitchers, and I. Walker. 2000. *Toolboxes for Maintaining and Improving Drinking Water Quality.* London, England: UK Water Industry Research Limited (UKWIR).

Characklis, W.G. 1990. Laboratory Biofilm Reactors. In *Biofilms*. Edited by W.G. Characklis, and K.C. Marshall. New York: John Wiley Publishers.

Clark, R.M. 1991. Water Quality Modeling in Distribution Systems: Overview and Conference Summary, In Proc. of AwwaRF and USEPA Conference on Water Quality Modelling in Distribution Systems. Cincinatti, Ohio, 4-5 February 1991. Denver: AWWA.

Clark, R.M. 1998. Chlorine Demand and TTHM Formation Kinetics: A Second- Order Model. *Journal of Environmental Engineering*, 124(1):16-24

Clark, R.M., and W.M. Grayman. 1998. *Modeling Water Quality in Drinking Water Distribution Systems*. Denver, Colo.: AWWA.

Clark, R.M., W.M. Grayman, J.A. Goodrich, R.A. Deininger, and A.F. Hess. 1991. Field-Testing Distribution Water Quality Models. *Journal of the American Water Works Association*, 183(7):67-75.

Clark, R.M., W.M. Grayman, and R.M. Males. 1988. Contaminant Propagation in Distribution Systems. *Journal of Environmental Engineering*, 114(4):929-943.

Clark, R.M., and M. Sivaganesan. 1998. Predicting Chlorine Residuals and the Formation of TTHMS in Drinking Water. *Journal of Environmental Engineering*, 124(12):1203-1210.

Clark, R.M., and M. Sivaganesan. 2002. Predicting Chlorine Residuals in Drinking Water: A Second Order Model. *Journal of Water Resources Planning and Management*. 128(2):1-10

Clark, R.M., R. Thurnau, M. Sivaganesan, and P. Ringhand. 2001. Predicting the Formation of Chlorinated and Brominated By-Products. *Journal of Environmental Engineering*. 127(6):493-501

Clark, R.M., G. Smalley, J.A. Goodrich, R. Tull, L.A. Rossman, J.J. Vasconcelos, and P.F. Boulos. 1994. Managing Water Quality in Distribution Systems: Simulating TTHM and Chlorine Residual. *Journal of Water Science Research and Technology - Aqua*, 43(4):182-191.

Cleasby, J.L., E.R. Baumann, A.H. Dharmarajah, and G.L. Sindt. 1989. *Design and Operation Guidelines for Optimization of the High-Rate Filtration Process*. Denver, Colo.: AwwaRF and AWWA.

Clement, J., M. Hayes, P. Sarin, V.L. Snoeyink, W.M. Kriven, J. Bebee, K. Jim, M. Beckett, G.J. Kirmeyer, and G. Pierson. 2002. *Development of Red Water Control Strategies*. Denver, Colo.: AwwaRF and AWWA.

Colbourne, J.S. 1985. Materials Usage and their Effects on the Microbiological Quality of Water Supplies. *J Appl. Bacteriol. Symp. Series,* 47S-59S.

Colbourne, J.S. and D.A. Brown. 1979. Dissolved Oxygen Utilisation as an Indicator of Total Microbial Activity on Non-Metallic Materials in Contact With Potable Water. *J. Bacteriol.* 47:223-231.

Colbourne, J.S., P.J. Dennis, W. Keevil, and C.W. Mackerness. 1991. The Operational Impact of Growth of Coliforms in London's Distribution System. In *Proc. of the AWWA Annual Water Quality Technology Conference, Orlando*. Denver, Colo.: AWWA.

Colin, F., G. Grapin, J. Cheron, Y. Levi, E. Pozzoli, J. Miazza, and O. Pascal. 1987. Etude de L'Évolution de la Qualité d'eau Potable Dans les Réseaux de Distribution, une Approche et des Moyens Nouveaux. *Techn. Sci. Munic. Eau.* 12:565-574.

Coulson, J.M., and J.F. Richardson. 1964. *Chemical Engineering*. Volume 2, 2nd edition. Oxford, England: Pergamon Press.

Creasey, J., Chambers, K. and Russel, S. 1997. *Measuring the Age of Water in Distribution. WRc Portfolio Report No. PT 2060.* Swindon, UK: WRc.

Deininger, R.A., R.M. Clark, A.F. Hess and E.V. Bernstam. 1992. Animation and Visualization of Water Quality in Distribution Systems. *Journal of the American Water Works Association,* 184 (7):48-52.

Delanoue, A., Holt, D.M., Wodward, C.A., McMath, S.M. and Smith, S.E. 1997. Effect of Pipe Materials on Biofilm Growth and Deposit Formation in Water Distribution Systems. In *Proc of the AWWA Annual Water Quality Technology Conference, Denver, 9-12th November 1997.* Denver, Colo.: AWWA.

Desjardins, A., Gauthier, V., Mainville, E., Millette, R. and Marche, C. 2000. Determination of Residence Time in Water Distribution Systems by Signal Processing. In *Proc. of the AWWA Annual Water Quality Technology Conference, Salt Lake City, November 5-9, 2000*. Denver, Colo. AWWA.

Dukan, S., Y. Levy, P. Piriou, F. Guyon and P. Villon. 1996. Dynamic Modelling of Bacterial Growth in Drinking Water Networks. *Water Research,* 30(9):1991-2002.

DWI (UK Drinking Water Inspectrorate). 2002. *Drinking Water 2001 (Annual Report).* London, UK: DWI. [Online]. Available: <http://www.dwi.gov.uk/pubs/annrep01>. [cited January 6, 2003]

Engineering Computer Applications Committee. 1999. Calibration Guidelines for Water Distribution System Modeling. In *Proc. of AWWA Information Management and Technology Conference, April 18-21, 1999, New Orleans, Louisiana.* Denver, Colo.: AWWA.

Edwards, D.K., V.E. Denny, and A.F. Mills. 1976. *Transport Processes*. New York: McGraw-Hill.

Elton, A., L.F. Brammer, and N.S. Tansley. 1995. Water Quality Modeling in Distribution Networks. *Journal of the American Water Works Association*, 187(7):44-52.

Fransolet, G., Villiers, G and Masschelein, W.J. 1985. Influence of Temperature on Bacterial Development in Waters. *Ozone: Science and Engineering*, 7(3):205-227.

Gagnon, G.A., P.J.Ollos, and P.M.Huck. 1997. Modelling BOM Utilisation and Biofilm Growth in Distribution Systems: Review and Identification of Research Needs. *Journal of Water Science Research and Technology - Aqua*, 46(1):165-180.

Gale, P. Lacey, R. Stanfield, G and Holt D. 1997. The Development and Application of Coliform Predictive Models for the UK Water Industry. *J. Water SRT - Aqua,* 46(4):185-195.

Gatel, D., M. Mercier, C. Volk, and J-C. Joret. 1995. Control of Bacterial Regrowth in Distribution Systems: Influence of BDOC and Chlorine Residual. *Water Supply*, 13(2):275-290.

Geldreich, E.E. 1991. Investigating the Outbreak in Cabool, Missouri for a Water Supply Connection. In *Proc. of AwwaRF and USEPA Conference on Water Quality Modeling in Distribution Systems. Cincinatti, Ohio, 4-5 February 1991*. Denver: AWWA.

Grayman, W.M. 2001. Use of Tracer Studies and Water Quality Models to Calibrate a Network Hydraulic Model. *Current Methods*. 1(1):38-42.

Grayman, W.M., R.M. Clark, and R.M. Males. 1988. Modeling Distribution-System Water Quality: Dynamic Approach. *Journal of Water Resources Planning and Management*, 114(3):295-312.

Grayman, W.M., L.A. Rossman, C. Arnold, R.A. Deininger, C. Smith, J.F. Smith, and R. Schnipke. 2000. *Water Quality Modeling of Distribution System Storage Facilities*. Denver, Colo.: AwwaRF and AWWA.

Grayman, W.M., L.A. Rossman, M.A. Gill, Y. Li, and D.E. Guastella. [N.d.] Measuring and Modeling Disinfectant Wall Demand in Metallic Pipes. Forthcoming.

Haas, C.N., and J. Joffe. 1994. Disinfection under Dynamic Conditions: Modification of Hom's Model for Decay. *Environmental Science and Technology*, 28:1367-1369.

Haas, C.N., J. Joffe, U. Anmangandla, J.G. Jacongelo, and M. Heath. 1996. Water Quality and Disinfection Kinetics. *Journal of the American Water Works Association*, 88(3):95-103.

Haas, C.N. and Karra S.B. (1984). Kinetics of Wastewater Chlorine Demand Exertion. *Journal of the Water Pollution Control Federation*, 56(2):170-173.

Hague, J., C.T. Ta, and M.J. Biggs. 1999. Small-scale Models for Validation of CFD in Water Applications. In *Water Industry Systems: Modelling and Optimization Applications, Vol 1.* Edited by D.A. Savic and G.A. Walters. Baldock, Hertfordshire, England: Research Studies press Ltd.

Hambsch, B., P. Werner und F.H. Frimmel. 1992. Bakterienvermehrungsmessungen in Aufbereiteten Wassern Verschiedener Herkunft. *Acta Hydrochim. Hydrobiol.* 20(1):9-14.

Holt, D.M., Gauthier, V, Merlet, N. and Block, J-C. 1998. Importance of Disinfection Demand of Materials for Maintaining Residuals in Drinking Water Distribution Systems. *Specialised Conference on Drinking Water Distribution With and Without Disinfectant Residuals.* IWSA Mulheim am Ruhr, September 1998.

Holt, D.M, Smith, S.E., and Colbourne, J.S. 1995. Devising Practical Operational Strategies to Control Biofilm Growth in London's Water Distribution System. In *Proc. of the Water Quality Technology Conference, New Orleans, 12-16th November 1995.* Denver, Colo.: AWWA.

Holt, D.M, Todd, R.D., Delanoue, A. and Colbourne, J.S. 1995. A Study of Nitrite Formation and Control in Chloraminated Distribution Systems. In *Proc.of the Water Quality Technology Conference, New Orleans, 12-16th November 1995*. Denver, Colo.: AWWA.

Holt, D.M., T. Woodward, T. Rachwal, A. Delanoue and J.S. Colbourne. 1994. A Pipe Based Experimental System to Study Factors Influencing Bacterial Growth in London's Water Mains. In *Proc. of the AWWA Annual Water Quality Technology Conference, San Francisco*. Denver, Colo.: AWWA.

Hua, F., J.R. West, R.A. Barker, and C.F. Forster. 1999. Modelling of Chlorine Decay in Municipal Water Supplies. *Water Research*. 33(12): 2735-2746.

Huck, P.M. 1990. Measurement of Biodegradable Organic Matter and Bacterial Growth Potential of Drinking Water. *Journal of the American Water Works Association*. 82:78-56.

Jackson, P.J., and I. Sheiham. 1980. *Calculation of Lead Solubility in Water. WRc Technical Report TR 152*. Swindon, UK: WRc.

Jadas-Hécart, A., A. El Moher, M. Stitou, P. Bouillot, and B. Legube. 1992. The Chlorine Demand of a Treated Water. *Water Research*. 26(8):1073-1084.

Joll, C., W. Simanjuntak, A. Heitz, I. Kristiana, and R. Kagi. 2002. Catchment Contributions to Disinfection By-Products in a Surface Water Source. In *Proc. of Enviro 2002 and IWA 3rd World Water Congress, April 7-12, 2002, Melbourne, Australia*. Artarmon, New South Wales, Australia: Australian Water Association.

Joret, J.C. and Y. Levy. 1986. Méthode Rapide D'Évaluation du Carbone Éliminable des Eaux par Voie Biologique. *Trib. Cebedeau*. 510(39):3-9.

Khiari, D., S. Barrett, R. Chinn, A. Bruchet, P. Piriou, L. Matia, F. Ventura, I. Suffet, T. Gittelman, and P. Leutweiler. 2001. *Distribution Generated Taste-and-Odor Phenomena*. Denver, Colo.: AwwaRF and AWWA.

Kippin, S.J., Peet, J.R., Marshall, J. S. and Marshall, J.M. 2001. Water Quality Impacts From Blending Multiple Water Types [Project #469]. AWWA Denver, Colo.

Kirmeyer, G.J., G.W. Foust, G.L. Pierson, J.J. Simmler, and M.W. LeChevallier. 1993. *Optimizing Chloramine Treatment*. Denver, Colo.: AwwaRF and AWWA.

Kirmeyer, G.J., M. Friedman, J. Clement, A. Sandvig, P.F. Noran, K.D. Martel, D. Smith, M. LeChevallier, C. Volk, E. Antoun, D. Hiltebrand, J. Dyksen, and R. Cushing. 2000. *Guidance Manual for Maintaing Distribution System Water Quality*. Denver, Colo.: AwwaRF and AWWA.

Kirmeyer, G.J., Kirby L., Murphy B.M., Noran P.F., Martel K.D., Lund T.W., Anderson J.L. and Medhurst R. 1999. *Maintaining Water Quality in Finished Water Storage Facilities*. Denver, Colo.: AwwaRF and AWWA.

Kirmeyer, G.J., L.H. Odell, J. Jacangelo, A. Wilczak and R. Wolfe. 1995. *Nitrification Occurrence and Control in Chloraminated Water Systems*. Denver, Colo.: AwwaRF and AWWA.

Kuch, A. 1988. Investigations of the Reduction and Reoxidation Kinetics of Iron (III) Oxide Scales Formed in Water. *Corrosion Science.* 28(3):221-231.

Laurent, P., P. Servais, M. Prevost, D. Gatel, and B.Clement. 1997. SANCHO Model of Microbiological Water Quality in Distribution Systems. *Journal of the American Water Works Association*, 89(7):92-103.

LeChevallier, M.W. 1990. Coliform Regrowth in Drinking Water: A Review. Journal of the American Water Works Association :74-86.

LeChevallier, M.W., C.D. Lowry, and R.G. Lee. 1990. Disinfecting Biofilms in a Model Distribution System. *Journal of the American Water Works Association* :111-123.

LeChevallier, M.W., W. Schulz, and R.G. Lee. 1991. Bacterial Nutrients in Drinking Water. *Applied and Environmental Microbiology*, 57(3):857-862.

LeChevallier, M.W., N. Shaw, L.A. Kaplan and T. Bott. 1993. Development of a Rapid Assimilable Organic Carbon Method for Water. *Applied and Environmental Microbiology,* 59(5):1526-1531.

Lee, S.H., J.T. O'Connor and S.K. Barnerji. 1980. Biologically Mediated Corrosion and its Effects on Water Quality in Distribution Systems. *Journal of the American Water Works Association.* 80:636-645.

Liou, C.P., and J.R. Kroon. 1987. Modeling the Propagation of Waterborne Substances in Networks. *Journal of the American Water Works Association, 79:54-58.*

Lu, C., P. Biswas, and R.M.Clark. 1995. Simultaneous Transport of Substrates, Disinfectants and Mircoorganisms in Water Pipes. *Water Research,* 29(3):881-894.

Lucena, F., J. Frias and F. Ribas. 1990. A New Dynamic Approach to the Determination of Biodegradable Dissolved Organic Carbon in Water. *Env. Technol.* 12:343-347.

Lund, V., and K. Ormerod. 1995. The Influence of Disinfection Processes on Biofilm Formation in Water Distributions Systems. *Water Research*, 29(4):1013-1021.

Mackerness, C.W., J.C. Colbourne, P.J. Dennis, A.J. Rachwal, and C.W. Keevil. 1992. Formation and Control of Coliform Biofilms in Drinking Water Distribution Systems. In *Microbial Biofilm Formation and Control, Society for Applied Bacteriology Technical Series 30.* Edited by S.P. Denyer, S.P. Gorman, and M. Sussman. London: M. Blackwell Scientific Publications.

Mackerness, C.W., J.C. Colbourne, and C.W. Keevil. 1991. Growth of *Aeromonas Hydrophila* and *Escherichia Coli* in a Distribution System Biofilm Model. In *Proc. of the UK Symposium on Health Related Water Microbiology, University of Glasgow, September 1-3, 1991.* London: IAWPRC.

MacRae, A.W., G. Keay, and M.S. Farrimond. 1993. Customer Satisfaction; Taste and Odour; Microbial Standards; are they Competing Objectives. *Water Supply*, 11(3/4):27-36.

Maier, S.H., Woodward, C.A., Delanoue, A., Holt, D.M, McMath, S.M. and Powell, R.S. 1999. Relationships Between Total Chlorine, Particle Counts and Biofilm Detachment in a Drinking Water Distribution System. In *Proc. of the Biofilms in Aquatic Environment Conference, Warwick, UK 13-16th April 1997.* Edited by C.W. Keevil, A Godfree, D. Holt and C. Dow. Royal Society of Chemistry. 231-238.

Males, R.M., W.M. Grayman, and R.M. Clark. 1988. Modelling Water Quality in Distribution Systems. *Journal of Water Resources Planning and Management*, 114(2):197-209.

Marshall, G.P. 2001. *Understanding and Preventing Discoloured Water.* London, England: UK Water Industry Research Limited (UKWIR).

Mary-Dile, V., D. Gatel, P. Servais, C. Elain, and J. Cavard. 2000. Developing a Water Quality Model for the Distribution System in the Suburbs of Paris, France. In *Proc. of the AWWA Annual Water Quality Technology Conference, Salt Lake City.* Denver, Colo.: AWWA.

Mathieu, L., J.L. Paquin, J-C. Block, G. Randon, G. Maillard, and D.J. Reasoner. 1992. Parameters Governing Bacterial Regrowth in Distribution Systems. [In French]. *J. Fr. Hydrol.*, 15(3):207-221.

McCoy, W.F. and J.W. Costerton. 1982. Growth of Sessile Sphaerotilus Natans in a Tubular Recycle System. *Applied and Environmental Microbiology*, 43:1490-1494.

McGhee, T.J. 1991. *Water Supply and Sewerage. 6th Edition.* Singapore: McGraw-Hill International.

McMath, S.M., Delanoue, A, Holt, D.M, Maier, S. and Woodward, C.A. 1999a. Importance of "Clumps" Shed From Biofilms in a Water Distribution System. In *Proc. of the Biofilms in Aquatic Environment Conference, Warwick, UK 13-16th April 1997.* Edited by C.W. Keevil, A Godfree, D. Holt and C. Dow. Royal Society of Chemistry. 72-80.

McMath, S. M., Sumpner, C, Holt, D.M., Delanoue, A. and Chamberlain, A.H.L. 1999b. The Fate of Environmental Coliforms in a Model Water Distribution System. *Letters in Applied Microbiology,* 28:93-97.

NanCIE (Centre International de l'Eau de Nancy). 1994. Propella. Patent 940738-F.

NanCIE (Centre International de l'Eau de Nancy). 1998. *Chlorine and Monochloramine Demand of Materials for Drinking Water Distribution Systems.* London, England: UK Water Industry Research Limited (UKWIR).

Nielson. 1994. International Report Denmark. *Water Supply.*12(1/2):IR7-IR15.

Norman, G., W.G. Characklis and J.D. Bryers. 1977. Control of Microbial Fouling in Circular Tubes With Chlorine. *Developments in Industrial Microbiology,* 18:581-590.

Parent, A., S. Fass, M.L. Dincher, D. Reasoner, D. Gatel, and J.C. Block. 1996. Control of Coliform Growth in Drinking Water Distribution Systems. *Journal of the Chartered Institute of Water and Environmental Management*, 10(12):442-445.

Pedersen, K. 1982. Method for Sudying Microbial Biofilms in Flowing-Water Systems. *Applied and Environmental Microbiology,* 43:6-13.

Pernitsky, D.J., G.R. Finch, and P.M. Huck. 1995. Disinfection Kinetics of Heterotrophic Plate Count Bacteria in Biologically Treated Potable Water. *Water Research*, 29(5):1235-1241.

Piriou, P. 1994. A New Tool for the Study of the Evolution of Water Quality in Distribution Systems: Design of a Network Pilot. In *Proc. of the AWWA Annual Water Quality Technology Conference, New York, June 19-23, 1994*. Denver, Colo.: AWWA.

Piriou, Ph., S. Dukan, Y.Levi, and P.A. Jarrige. 1997. Prevention of Bacterial Growth in Drinking Water Distribution Systems. *Water Science and Technology*, 35(11-12):283-287.

Piriou, P. and Y. Lévi. 1994. A New Tool for the Study of the Evolution of Water Quality in Distribution Systems: Design of a Network Pilot. In *Proc. of the AWWA Annual Water Quality Technology Conference, New York, June 19-23, 1994.* Denver, Colo.: AWWA.

Powell, J.C. 1998. Modelling Chlorine in Water Distribution Networks. Ph.D. thesis., The University of Birmingham, Birmingham, England.

Powell, J.C., N.B. Hallam, J.R. West, C.F. Forster and J.Simms. 2000b. Factors Which Control Bulk Chlorine Decay Rates. *Water Research*, 34(1):117-126.

Powell, J.C., J.R. West, N.B. Hallam, C.F. Forster and J.Simms. 2000a. Performance of Various Kinetic Models for Chlorine Decay. *Journal of Water Resources Planning and Management*, 126(1):13-20.

Powell, J.C., J.R. West, N.B. Hallam, and J.Simms. 1999. Durability and Maintenance of Chlorine Network Models. In *Proc. of AWWA Information Management and Technology Conference, April 18-21, 1999, New Orleans, Louisiana.* Denver, Colo.: AWWA.

Prakash, T.M., R.J. Scholze, C.H. Neff, S.W. Maloney, M. Heath and E.D. Smith. 1988. *Development of the Pipe Loop System Determining Effectiveness of Corrosion Control Chemicals in Potable Water Systems.* Technical Report N-88/12. US-CERL.

Ransome, M.E., T.N. Whitmore, and J. Easton. 1995. Practical Implications of Disinfection. *Water Supply*, 13(2):131-142.

Reddy, L.S., L.E. Ormsbee, and D.J. Wood. 1995. Time-Averaging Water Quality Assessment. *Journal of the American Water Works Association*, 187 (7):64-73.

Rodriguez, M.J., and Sérodes J.B. 1994. Development of Neural Net-Based Models for Water Quality Managemnet and Prediction in Distribution Systems. In *Proc. of AWWA conference on computers in the water industry. Los Angeles, California, 10-13 April 1994*. Denver: AWWA.

Rossman, L.A. and P.F. Boulos. 1996. Numerical Methods for Modeling Water Quality in Distribution Systems: a Comparison. *Journal of Water Resources Planning and Management*, 122(2):137-146.

Rossman, L.A., P.F. Boulos, and T. Altman. 1993. Discrete Volume-Element method for Network Water-Quality Models. *Journal of Water Resources Planning and Management*, 119(5):505-517.

Rothberg, Tamburini and Winsor Water Softening Calculation Model, Version 2000. Software available from AWWA. Denver.

Sarin, P., V.L. Snoeyink, J. Bebee, W.M. Kriven, and J.A. Clement. 2001. Physico-Chemical Characteristics of Corrosion Scales in Old Iron Pipes. *Water Research*. 35(12)2961-2969.

Seidal, C., M.M. Frey, R. Hund, N. Graziano, B. Takeda, B. Wischer, and J. McEnroe. 2002. The IDSE: an Alternative Approach Using Distribution System Models. In *Proc. of 2002 AWWA Annual Conference*. Denver, Colo.: AWWA.

Servais, P., Laurent P., Billen G., and Gatel D. 1995. Development of a Model of BDOC and Bacterial Biomass Fluctuations in Distribution Systems. *Revue des Sciences de L'Eau*, 8(4):427-462.

Servais, P., G. Billen and M.C. Hascoët. 1987. Determination of the Biodegradable Fraction of Dissolved Organic Matter in Waters, *Water Research*, 21(4):445-450.

Singer, P. 1999. Humic Substances as Precursors for Potentially Harmful Disinfection By-Products. *Water Science and Technology*. 40(9):25-30.

Skipworth, P.J., Machell, J. and Saul, A.J. 2002. Empirical Travel Time Estimation in a Distribution Network. *Water and Maritime Engineers*, 154(1):41-49.

Smith, S.E., T. Ta, D.M. Holt, A.Delanoue,and J.S. Colbourne. 1998. Minimising Red Water in Drinking Water Distribution Systems. In *Proc. of the AWWA Water Quality Technology Conference, November 1-4, 1998, San Diego*. Denver, Colo.: AWWA.

Smith, S.E., T. Ta, D.M. Holt, A. Delanoue, J.S. Colbourne, A.H.L. Chamberlain, and B.J. Lloyd. 1999. A Pipeline Facility for the Examination of Pipe Wall Deposits and Red Water Events in Drinking Water. *Journal of the Chartered Institution of Water and Environmental Management*. 13(10):7-15.

Sohn, J., Amy, G., Boireau, A., Dourdin, G., Gatel, D. and J Cavard. 1998. In-Plant Versus Distribution Formation of Chlorination By-Products. In *Proc. Water Quality Technology Conference of the American Water Works Association, San Diego, Ca*. Denver, Colo.: AWWA.

Spiering, J.W.F. 2002. *Drinkwater Onderzoek Verblijftijden in Aansluitleidingen*. Report No. 202451. Netherlands: AZT (Advisering and Zuiveringstechnologie).

Stanfield, G. and P.H. Jago. 1987. *The Development and Use of a Method for Measuring the Concentration of Assimilable Organic Carbon in Water.* WRc-report PRU 1628-M. Medmenham, UK: WRc.

Teefy, S. 1996. *Tracer Studies in Water Treatment Facilities: A Protocol and Case Studies.* Denver, Colo.: AwwaRF and AWWA.

Twort, A.C., D.D. Ratnayaka, and M.J. Brandt. 2000. *Water Supply.* 5th ed. London: IWA.

USEPA (US Environmental Protection Agency). 2002. EPANET 2.0: A water distribution hydraulic and water quality simulation program for personal computers. [Online]. Available: <http://www.epa.gov/ORD/NRMRL/wswrd/epanet.html> [cited January 6, 2003].

Valentine, R.L., K. Ozekin, and P.J. Vikesland. 1998. *Chloramine Decomposition in Distribution Systems and Model Waters.* Denver, Colo.: AwwaRF and AWWA.

Valentine, R.L. and P.J. Vikesland. 2000. *Role of the Pipe-Water Interface in Distinfectant By-Product Formation and Disinfectant Loss.* Denver, Colo.: AwwaRF and AWWA.

van den Boomen, M., and J.H.G. Vreeburg. 1999. *Nieuwe Ontwerprichtlijnen Voor Distrbutienetten,* Kiwa report No. SWE 99.011. Nieuwegiein, Netherlands: Kiwa.

van den Hoven, T., D. van der Kooij, J. Vreeburg, and H. Brink.1994. Methods to Analyse and to Cure Water Quality Problems in Distribution Systems. *Water Supply,* 12 (3/4):151-159.

van der Kooij, D. 1992. Assimilable Organic Carbon as an Indicator of Bacterial Regrowth. *Journal of the American Water Works Association*, 84(2):57-65.

van der Kooij, D., Y.C. Drost, W.A.M. Hijnen, J. Willemsen-Zwaagstra, P.J. Nobel, and J.A. Schellart. 1995. Multiple Barriers Against Micro-Organisms in Water Treatment and Distribution in the Netherlands. *Water Supply*, 13(2):13-23.

van der Kooij, D., J.H.M. van Lieverloo, P. Gale, and G. Stanfield. 2002. *Distributing Drinking Water With low or Zero Disinfectant Residual.* London, England: UK Water Industry Research Limited (UKWIR).

van der Kooij, D., J.H.M. van Lieverloo, J.A. Schellart and P. Hiemstra. 1999. Maintaining Quality without Disinfectant Residual. *Journal of the American Water Works Association*, 91(1):55-64.

van der Kooij, D. and H.R. Veenendaal. 1993. Assessment of the Biofilm Formation Characteristics of Drinking Water. In *Proc. American Water Works Association Water Quality Technology Conference, Toronto Ontario, 1992.* Denver, Colo.: AWWA.

van der Kooij, D., A. Visser, and W.A.M. Hijnen. 1982. Determining the Concentration of Easily Assimilable Organic Carbon in Drinking Water. *Journal of the American Water Works Association*, 74(10):540-545.

Vasconcelos, J.J., L.A. Rossman, W.A. Grayman, P.F. Boulos, and R.M. Clark. 1997 Kinetics of Chlorine Decay. *Journal of the American Water Works Association,* 89(7):54-65.

Vasconcelos, J.J., P.F. Boulos, W.M. Grayman, L. Kiene, O. Wable, P. Biswas, A. Bhari, L.A. Rossman, R.M. Clark, and J.A. Goodrich. 1996. *Characterization and Modeling of Chlorine Decay in Distribution Systems.* Denver, Colo.: AwwaRF and AWWA.

Volk, C and Joret, J.C. 1994. Parametres Predictiifs de L'Apparition des Coliformes Dans le Reseaux de Distribution d'eau D'Alimentation. *Rev. Sci de l'eau,* 7:145-166.

Walski, T., D.V. Chase, and D. Savic. 2001. *Water Distribution Modeling.* Waterbury, CT.: Haestad Press.

Walski, T., D.V. Chase, D. Savic, W.M. Grayman, S. Beckwith, and E. Koelle. 2003. *Advanced Water Distribution Modeling and Management.* Waterbury, CT.: Haestad Press.

Werner, P. 1985. Eine Methode zur Bestimmumng der Verkeimungsneigung von Trinkwasser. *Vom Wasser,* 65:257-270.

Westerhoff, P., D. Reckhow, G. Amy, and Z. Chowdhury. 2002. *Mechanistic-Based Disinfectant and Disinfectant By-Product Models, USEPA Grant # R 826831-01-0, Final Report.*[Online]. Available: <http://ceaspub.eas.asu.edu/pwest/USEPA-CL2-DBP-2002.htm> [cited January 6, 2003].

WHO (World Health Organisation). 1993. *World Health Organisation Guidelines for drinking water quality, Volume I,* 2nd Edition. Geneva: WHO.

WHO (World Health Organisation). 1996. *World Health Organisation Guidelines for drinking water quality, Volume II,* 2nd Edition. Geneva: WHO.

WHO (World Health Organisation). 1997. *World Health Organisation Guidelines for drinking water quality, Volume III.* Geneva: WHO.

Williams, A.M., Andrews, S and Wakeford, P. 2001. The Control of Nitrite Formation in London's Distribution System. *J. CIWEM* 15:162-166.

Woodward, C.A., Ta, C.T., Colbourne, J.S and Holt, D.M. 1995. Behaviour of Particles in a Large Scale Experimental Pipe System. In *Proc. of the Water Quality Technology Conference, New Orleans, 12-16th November 1995.* Denver, Colo.: AWWA.

Woodward, C.A., C.T. Ta, D. Holt and J. Colbourne. 1996. Relationships Between Observed Monochloramine Decay Rates and Other Physical and Chemical Parameters in a Large Scale Experimental Pipe System. Presented at meeting of Institution of Water and Environmental Management South Eastern Branch, Redhill, Surrey, England, 17 January 1996.

WRc (Water Research Centre). 1989. *Network Analysis – A Code of Practice*, Swindon, England: WRc.

WRc (Water Research Centre). 1997a. *Changes in The Quality of Treated Water During Storage.* London, England: UK Water Industry Research Limited (UKWIR).

WRc (Water Research Centre). 1997b. *Predictive Statistical Analysis of Coliform Occurrences.* London, England: UK Water Industry Research Limited (UKWIR).

Young, W.F., H. Horth, R. Crane, T. Ogden, and M. Arnott. 1996. Taste and Odour Threshold Concentrations of Potential Potable Water Contaminants. *Water Research*, 30(2):331-340.

ABBREVIATIONS

A	cross-sectional area
ANN	artificial neural network
AOC	assimilable organic carbon
ATP	adenosine triphosphate
AwwaRF	Awwa Research Foundation
AWWA	American Water Works Association
BDOC	biodegradable organic carbon
BFR	biofilm formation rate
$[Br]$	bromine concentration
°C	degrees Celsius
C	concentration
C_0	initial concentration
CAD	computer aided design
CFD	computational fluid dynamics
cfu	colony forming unit
cfu/mL	colony forming unit per milliliter
CFX	CFX – CFD modeling software company
C_L	limiting concentration
$[Cl_2]$	chlorine dose
$[Cl_2]_0$	initial chlorine dose
$[Cl_2]_1$	new chlorine dose
cm^2	square centimeter

C_t	concentration at time t
CT	product of concentration and contact time
d	internal diameter
d^{-1}	per day
DBP	disinfection by-products
DBPR	Disinfectant and Disinfectant By-Product Rule
DIC	dissolved inorganic carbon
DMA	district meter area
DOC	dissolved organic carbon
E	activation energy
EU	European Union
EU std	European Union standard
F	frequency factor
f{}	function of parameter enclosed in brackets
ft	foot
ft/s	foot per second
GAC	granular activated carbon
GIS	geographical information system
GLM	generalized linear modeling
h	hour

h^{-1}	per hour
HAA	haloacetic acids
HPC	heterotrophic plate count
IDSE	initial distribution system evaluation
in.	inch
J/mol	joule per mole
J/mol °C	joule per mole and degree Celsius
k	reaction rate constant
k^*	pseudo first order reaction rate constant
k_b	bulk water decay coefficient
k_f	mass transfer coefficient
km	kilometers
k_{vcsc}	Valentine chloramine stability index
k_w	wall reactivity coefficient
k_{w*}	wall decay coefficient
L	length
LCR	Lead and Copper Rule
LRAA	local running annual average
L/s	liter per second
λ	Darcy-Weisbach friction factor

m	Hom disinfection kinetic parameter (variable units)
MDPE	medium density polyethylene
mg	milligrams
mg C/L	milligrams carbon per liter
mg $CaCO_3$/L	milligrams as $CaCO_3$ per liter
mg/L	milligrams per liter
ML	megaliter
mL	milliliter
mm	millimeter
mol	mole
μg C/L	micrograms carbon per liter
μg/L	micrograms per liter
n	reaction order
n'	coefficient of dilution
N	number of surviving microorganisms at time t
N_0	number of surviving microorganisms at time zero
NanCIE	Centre International de l'Eau de Nancy
N.d.	no date
NDMA	N-nitrosodimethylamine
NOM	natural organic matter
NPDWR	national primary drinking water regulations
$[NH_2Cl]$	monochloramine concentration at time t
$[NH_2Cl]_0$	initial monochloramine concentration
$[Nitrate]$	nitrate (NO_3) concentration

[*Nitrite*]	nitrite (NO_2) concentration
PAH	polynuclear aromatic hydrocarbon
PIV	particle image velocimetry
PVC	polyvinyl chloride
Q	flow
R	ideal gas constant (8.31 J/mol°C)
R2A	Reasoner and Geldreich agar
s	second
SCADA	supervisory control and data acquisition
SDS	simulated distribution system
SMP	standard monitoring program
SSS	system specific study
SWTR	Surface Water Treatment Rule
T	temperature (°C)
t	time
TCR	Total Coliform Rule
THM	trihalomethanes
[THM]	THM concentration at time t
$[THM]_0$	THM concentration given by initial chlorine dose
$[THM]_1$	THM concentration given by new chlorine dose

TOC	total organic carbon
[*TOC*]	TOC concentration
TWL	top water level
u	velocity
UK	United Kingdom
UK std	United Kingdom standard regulations
UKWIR	United Kingdom Water Industry Research Ltd
USA	United States of America
USEPA	United States Environmental Protection Agency
UV	ultraviolet
[*UV*]	ultraviolet light concentration
V	volume
w	wall reaction pipe roughness coefficient
WHO	World Health Organization
WRc	Water Research Centre Ltd.
WTP	water treatment plant
X_0 to X_7	WTP model fitted coefficients

IWA Publishing's authorised EU representative for General Product Safety Regulations is Diane D'Arras, 15 rue Duret, 75116 Paris, France, e-mail: safety@iwap.co.uk.

Printed and bound by CPI Group (UK) Ltd, Croydon, CR0 4YY
12/05/2026
02108874-0013